D0962236

HEAD STRONG

ALSO BY DAVE ASPREY

The Bulletproof Diet

Bulletproof: The Cookbook

The Better Baby Book

HEAD STRONG

THE BULLETPROOF PLAN TO ACTIVATE UNTAPPED BRAIN ENERGY TO WORK SMARTER AND THINK FASTER—IN JUST TWO WEEKS

DAVE ASPREY

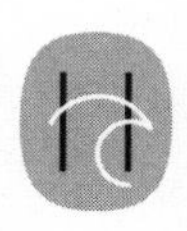

HARPER WAVE
An Imprint of HarperCollins*Publishers*

This book contains advice and information relating to health care. It should be used to supplement rather than replace the advice of your doctor or another trained health professional. If you know or suspect you have a health problem, it is recommended that you seek your physician's advice before embarking on any medical program or treatment. All efforts have been made to assure the accuracy of the information contained in this book as of the date of publication. This publisher and the author disclaim liability for any medical outcomes that may occur as a result of applying the methods suggested in this book.

FIRST EDITION

Library of Congress Cataloging-in-Publication Data

Asprey, Dave.

Head strong : the bulletproof plan to activate untapped brain energy to work smarter and think faster—in just two weeks / Dave Asprey.

ISBN 978-0-06-265241-6

1. Mental health—Nutritional aspects. 2. Brain—Care and hygiene. 3. Brain—Diseases—Prevention.

RC455.4.N8 A87 2016

616.8—dc23 2016047849

17 18 19 20 21 RRD 10 9 8 7 6 5 4 3 2 1

To you, dear reader, at the first time you hit the wall,
when you reach deep within yourself only to find that
there is no reservoir of energy and willpower waiting for you.
This book will help you move that wall out of your way.

CONTENTS

INTRODUCTION

If you're like most people, you learned the same things I did growing up: your intelligence is fixed, and your performance is a matter of effort. If you're stupid, you're stupid, and there's nothing you can do about it. If you fail, it's because you must be lazy and didn't try hard enough . . . or maybe you're just not that strong. Next time, you'll have to summon even more willpower to work harder, but if you fail again it's because *you* are a failure. It's your fault.

The belief that we either need to have superhuman smarts or put in an incredible amount of effort in order to be successful is built into our culture. We worship the struggle and the idea of "natural" talent. *Trying hard* is how you win. *Being smart* is how you win. But what if it doesn't have to be that way? What if it can be easier?

In my experience, fear of failure could drive an amazing amount of achievement. Before I was thirty, I was fortunate enough to have already attained the kind of career that would have made someone happy who was decades older than I was. I'd run technology strategy for a company valued at $36 billion, played a meaningful role in creating the infrastructure of the modern Internet, and sat on the highest-level advisory board of IBM (International Business Machines). I'd already made (and lost) $6 million. I was getting my MBA at Wharton, one of the top Ivy League business schools in the world, while working full-time as an executive at a start-up that eventually sold for more than $600 million. In short: I kicked ass, and it looked easy.

But behind that success was a constant hidden struggle. The people around me could see my physical challenges, but most had no idea about the level of inner struggle it took to perform. For example, it was obvious that I was overweight and out of shape, and I famously fell asleep in meetings. But few people knew that it was a battle for me just to get through the day because my brain wasn't working the way I needed it to. I couldn't focus at work, I had trouble retaining new information, and I had started to feel a chronic, debilitating fatigue that couldn't be explained away by an entrepreneur's lack of sleep.

I felt foggy, as if I had a constant hangover—it seemed like something was broken in my brain. I was irritable and cranky, quick to anger, and I made impulsive decisions. But I kept pushing myself. My middle finger got an excellent workout thanks to my driving habits, but the rest of my body was bloated and out of shape. It felt like I had to work twice as hard as other people to get the same amount done. I was pushing my body's accelerator all the way to the floor yet was somehow stalled in neutral, idling. Life felt hard because I *knew* I had more to give, but it just wouldn't come when I asked for it.

Then I started to fail some classes at Wharton despite trying my best. If only I had more willpower. If only I were smarter. I got really worried about flunking out of my MBA program, so I kept working harder and harder with no results. I wondered if my classmates were all smarter than I was. I didn't understand why I wasn't getting a better outcome, no matter how much effort I put in, and concluded that despite my success, I simply wasn't as good as I had thought.

What I didn't know then was that my fatigue, lack of focus, forgetfulness, moodiness, and even cravings weren't my fault. I wasn't lazy, a bad person, or a failure. The problem was that my brain was literally losing energy and wouldn't perform to the standard I wanted, no matter how hard I tried. Pushing the accelerator on a car with a broken engine won't make it go faster, no matter how hard you press.

Frustrated and afraid that I was going to lose everything I'd worked so hard to achieve, I began looking for a way to apply my computer hacking skills to these problems. I was lucky to find Dr. Daniel Amen's breakthrough book, *Change Your Brain, Change Your Life,* which mentioned something called a SPECT (single-photon emission computed tomography) scan, a nuclear imaging test that tells you how each part of your brain is using energy. At the time, this test was controversial and there were plenty of skeptics, but I was desperate and intrigued, so I went to try it out at Silicon Valley Brain Imaging. The nurse injected radioactive sugar into my arm and asked me to try to concentrate while they looked at the activity in my brain using a giant machine like an MRI (magnetic resonance imaging).

Lo and behold, the scan showed that my prefrontal cortex—the most highly evolved part of the brain that manages complex cognitive behavior and decision-making—had very little metabolic activity and was creating almost no energy. When I tried to focus and think, the part of my brain that was supposed to jump into action simply didn't show any signs of life. The psychiatrist took one look at the scan and said the words I will never forget: "Dave, inside your brain is total chaos. I have no idea how you're standing in front of me right now. You have the best camouflage I've ever seen." He was the first person to recognize the incredible effort I was applying just to function, and it was because he could actually see with his own eyes that my brain wasn't creating and using energy the way it should.

It wasn't exactly the best news I'd ever heard, yet that knowledge set me free. Suddenly I realized that I wasn't struggling to succeed because I was a personal failure or because maybe I'd bitten off more than I could chew. There was something physically wrong with my brain, and it was harming my performance. That day, my diminished brain function stopped being a moral problem and became an issue with my body's hardware that I could fix. I just had to figure out the causes of my system's weaknesses so I could eliminate them. As a career computer security technologist (a.k.a. hacker), that's exactly what I did for a living—take control of complex systems. And just like that, an idea was born: I would hack my brain to maximize its performance. You don't have to know everything about a system to hack it!

I've spent seventeen years and more than $1 million on a journey to uncover the secrets to creating a state of high performance, power, and resilience. I started using "smart drugs" (cognitive enhancers), which I've now benefited from for nearly twenty years, and I used the famous "real-life *Limitless* pill," modafinil, to turn my brain back on so I could finish my MBA at Wharton while working full-time. (Smart drugs, or "nootropics," are not considered academic doping, or else I'd have to give my degree back!)

I poured the extra energy and performance I got from the smart drugs into experimenting with and gaining an understanding of every conceivable method that might upgrade my brain's performance. I tried oxygen masks, cerebral lasers, brain-training software, EEG (electroencephalogram) neurofeedback (a type of biofeedback that uses real-time displays of brain activity), breathing exercises, electrical stimulation, ice baths, yoga, meditation, diets, pharmaceuticals, hormones, and every possible supplement—all to figure out what worked, what didn't, and why. I even became president and chairman of a twenty-three-year-old anti-aging nonprofit in Palo Alto called the Silicon Valley Health Institute so I could spend more time with experts in each of those fields.

Instead of allowing my brain to respond wildly to the stressors around me, I took control of my nervous system stress response and the very energy production in my cells in order to grow my resilience. The things I learned along the way were nothing short of life changing. After a lot of experimenting, I discovered that relatively easy lifestyle changes could lead to much higher brain energy. This gave me the power to focus even in the most distracting environments, to remember more., and to think more creatively.

The fatigue that once caused me to drag myself through the day was replaced with sustained energy and resilience to stress. My chronic sinus problems disappeared and the crippling jet lag that used to plague me for days when I traveled no longer affected me. I gained extra time to be productive because I was more effective, and at the end of the day I still had the energy and willpower to be grateful and full of joy. The extra energy also gave me the ability to focus on personal development and to dig really deep into EEG neurofeedback so I could understand and change my brain waves. All of these results added up to a level of performance that I never knew I was capable of. And it made me a better, happier human being.

Best of all, it was effortless. It almost didn't seem fair that my life had become so much easier than I had learned to expect it to be. My brain was so sharp and full of energy that I stopped taking modafinil after eight years—I just didn't need it anymore. But the greatest proof of my transformation came when I went back to Dr. Amen for a follow-up scan a decade after my first visit, and it revealed a high-functioning brain without the gaps I'd had years before. I was still working hard, but I had stopped struggling. I have come to believe that high-performance brains are our birthright as humans.

Since then, I've used the same techniques that worked for me with hundreds of thousands of followers and coaching clients. These are people of all ages and backgrounds—students, teachers, busy parents, and chief executive officers of Fortune 500 companies. Some are already performing at high levels and want to reach new heights, but the majority are facing some amount of fatigue and low brain energy like I was. No matter their current level of performance, all of these people have two things in common: they're willing to make the necessary changes to get the outcome they want, and they get results quickly by using my Head Strong techniques.

As a result of seeing one life-changing transformation after another, I decided to put the most effective practices together into a comprehensive two-week program designed to give you the fastest and most powerful results possible. It's difficult to boil tens of thousands of hours of research down to only the most important, easy-to-follow takeaways, and that's the challenge of creating a book like this. But my reason for writing this book is simple: the world is a better place when people are free to perform without the weight of exhaustion and wasted effort around their necks. And that's what I want for you. I wish someone had given me this book when I was failing out of Wharton or when I was an angry nineteen-year-old growing stretch marks.

If we injected radioactive sugar into your brain, do you think it would look perfect? It's unlikely, but if you're like most people, it would probably look, well . . . average. This book isn't about how to perform at an "okay" level. If you're comfortable with your daily routine, save yourself a few hours by giving this book to one of your friends who will actually use it.

Head Strong is for people who are looking for something more. This book is for you if you want to learn how to reach your maximum potential so that you can do whatever you love to do, do it better, and do it with less effort. It's for people who want to gain an edge, who want to spend more time enjoying the fruits of their labor and less time sweating it out in the field. If you're open to making a few simple changes in order to turn up the energy in your brain—so that you can do more, and be more—read on!

By following this program, you'll be able to upgrade your brain in just two weeks, and the results keep building from there. That means you are only fourteen days away from knowing what it feels like to have less stress, more focus, and more resilience than you've likely ever experienced. No matter who you are or where you're starting from, I know that you will be able to squeeze a significantly greater performance out of your brain. Everything you do will require less effort, and you will become a better, more highly performing version of yourself. What would it feel like to have maximum energy to fuel your brain, your cells, and your every desire? To be more patient with the people you love, to make the best possible decisions, and to enjoy every minute of it?

There's a name for this state where your body, mind, and emotions work together effortlessly to help you perform at levels beyond what you've learned to expect. I call it being Bulletproof, and I named my company after that high-performance state. And if you really want the energy to "bring it" every day, it's time to become Head Strong.

PART I

IT'S ALL IN YOUR HEAD

There are dozens of great books out there about brain health, and medical professionals who are my friends wrote many of them. *Head Strong* is not one of those books.

Sure, you want a brain free of disease. We all do. But what if you want more? What if your goal is not only to have a healthy brain but also to have a high-performing brain that works better right now—and lasts longer—than nature ever intended?

For years, I quietly upgraded every facet of my brain I could think of and spent hundreds of thousands of dollars to do it. Then, a few years ago, I went public with the concept of biohacking—taking control of your own biology so you can make your body do what you want it to do. The things I learned and the changes I made through biohacking improved my life in all sorts of ways, but what matters most to me is making my brain work as well as possible right now so that I can use it to enjoy my family, do good work, and have a positive impact on the world. And, of course, I want it to keep improving over time and last forever!

With Alzheimer's and other degenerative brain diseases on the rise, there is a lot of chatter out there about brain health. Our well-meaning health experts tell us to do crossword puzzles and go square dancing to keep our brains healthy. And that's all fine and good. But there is a major factor missing from these discussions. Before a brain actually becomes unhealthy or diseased, there are decades of only so-so brain performance.

Let me explain. When I started going to doctors years ago and told them that I wanted a higher level of mental performance, they assured me that my brain was perfectly healthy, chalked up my fatigue and lack of focus to stress, and sent me on my way. But what's the point of a "healthy" brain if it isn't working very well? Would you want to drive a car with an engine that ran only half as well as it was designed to?

A lot of the outdated discussions around brain health are based on the idea that we're only as good as the brain we're born with. Either you're sharp, smart, focused, and blessed with a good memory and the ability to learn new information easily—or you're not. It wasn't until later in the twentieth century that scientists discovered what's called neuroplasticity—the brain's ability to grow new cells and forge new neural connections throughout your life. Before then, researchers believed that your brain remained static until it degenerated in old age. (There are still practicing physicians today who were taught that in medical school!)

This is why the majority of the brain advice that we hear is focused on helping us avoid degeneration. Common advice hasn't caught up with the most recent brain research coming out of medical schools and neuroscience laboratories—but I have. Seriously. It has been my passion for years, and now as the founder of Bulletproof, I create nootropics (cognitive-enhancing substances) and run a thriving neuroscience institute, called 40 Years of Zen, that trains high-performance CEO brains, including my own.

You have a choice: you can wait a generation before this information hits the mainstream or you can benefit now. Neuroplasticity and new advances in cellular biology give you the ability to set up your brain for maximum performance by increasing energy production and new connections and reducing inflammation. This is a game changer. Before I upgraded my brain, I didn't realize how much my brain's performance affected every area of my life. Of course, I knew that my brain controlled my thoughts and my conscious mind, but I had no idea that it also controlled my relationships, my moods, my energy level, and even my food cravings.

The fact is, it's *all* in your head. How well your brain creates energy dictates how you will manage your every conscious and unconscious impulse, urge, decision, and desire. Your brain runs the operating system for your entire life, and it's time for a major upgrade.

1

HEAD START

YOUR BRAIN ON ENERGY

Think about your smartphone. When you first took it out of the box, it was so fast and efficient, wasn't it? It held its charge for a long time. It performed at its peak. Then you started downloading apps and filling up its memory with pictures and videos. The operating system grew increasingly bloated, and it stopped performing as well. Now it's slower to respond, and the battery is more quickly depleted. Nothing on your phone works as well as it did when it was new.

Your brain is not dissimilar, but instead of being clogged with selfies and cat videos, it gets drained by things in your diet and your environment that shouldn't be there. When most people think of toxins, they think of poison. And certainly, there are some poisonous chemicals that inhibit brain function: neurotoxins destroy or damage brain cells and weaken the body's ability to produce energy in cells.

But there are other, less frequently discussed things that I like to think of as brain kryptonite. These aren't just chemicals. Brain kryptonite includes anything that pulls needed energy away from the brain and into another part of the body. Certain foods, products in our environment, types of light, and even forms of exercise can weaken your brain. This brain kryptonite doesn't kill you—at least not at first—it just slowly and stealthily eats away at your battery life.

Your brain needs a lot of energy to perform well—in fact, the brain uses up to 20 percent of your body's overall energy.[1] That's more than any other organ in your entire body! So where does it get this energy? Your body makes it. Inside almost every cell in your body are at least several hundred tiny descendants of bacteria called mitochondria. The energy that sustains us is created in these mitochondria, and you'd be amazed at how important they are to the quality of your life. If your mitochondria stopped making energy in all your cells for even a few seconds, you'd die. The number, efficiency, and strength of your mitochondria dictates whether or not you'll eventually develop cancer or a degenerative disease, as well as how much brainpower you have right now. Who would have thought that these tiny organelles (organs inside each cell) were the key to your brainpower?

The body is amazingly efficient at producing energy and delivering it to exactly where it's needed, but any cell in the body is only able to store a few seconds' worth of energy at any given time. The body has to constantly make energy on demand, and from one moment to the next, it has no way of knowing what that demand is going to be. When you go to a job interview, your cells don't know ahead of time if the office is going to have fluorescent lighting that can slowly drain your mitochondria's energy. Suddenly, your brain is wasting some of its available energy to filter out that junk light, and you're left stumbling over your sentences and grasping for words. Your mitochondria can't keep up with your brain's energy demands.

Luckily for you, the prefrontal cortex—the "higher" part of your brain in charge of advanced cognitive function—has the most densely packed mitochondria of any part of your body (except for the ovaries!). That means that your mitochondria contribute more energy to your brain's performance than your heart, lungs, or legs. Your brain gets first dibs on mitochondrial energy, and your eyes and heart are right behind it in line.

When your body has to contend with toxins or brain kryptonite, or if it isn't creating and delivering energy as efficiently as it could, the body's demand for energy can exceed the supply. In these instances, you get mitochondrial energy "brownout" in parts of your body. The first symptom that your mitochondria are overtaxed is fatigue. Fatigue is an absolute performance killer. It causes cravings, moodiness, brain fog, forgetfulness, and lack of focus. Yes, most of the things you hate about yourself can stem from brain fatigue. It's not a moral failing. It's an energy delivery problem. When you have limitless energy, you stop needing to try hard to be a good person. You can learn to do it effortlessly, because it's what you always would have done if there were nothing in the way.

Your body has to make extra energy to get rid of toxins. This means that if toxins are draining your energy, your body becomes less and less efficient at metabolizing and removing them, and you'll have to expend even more energy to get rid of them. It's a vicious cycle that can wreck your performance if you don't do something to stop it.

Of course, this doesn't happen all at once, which is a good thing. If it did, you'd die. (Some fast-acting poisons like cyanide actually work by quickly stopping mitochondria.) The energy drain we deal with on a daily basis is a classic case of death by a thousand cuts. We live in an increasingly toxic world, and most of us eat toxic food. Our lifestyles (including the very same technology that makes us so efficient) also deplete our cellular energy reserves. Each one of these elements takes a little more energy away from your brain and away from your life.

Imagine that you're Superman (or Superwoman). One day, Lex Luthor pulverizes some kryptonite and sprinkles just a little bit of it around your house. If you eat (or inhale) a small amount of kryptonite dust, it won't kill you. You'll still be able to push through the day and save people, you'll just feel slightly off. In fact, you'll get used to feeling that way and believe it's normal. But as you keep ingesting a little bit more kryptonite every day, your ability to help people will slowly, invisibly decline until your body reaches the point where it's spending all of its energy trying to overcome the effects of the poison.

If you're anything like I used to be, you probably think that these symptoms of brain weakness are natural or perhaps just an unavoidable part of getting older. That's because almost everyone has some of these symptoms, which medicine defines as normal or "healthy." That's why normal is your nemesis—it's considered "normal" to grow increasingly tired and foggy as you age until one day you wake up with dementia, unable to remember the things that matter most.

Screw that noise.

Wouldn't you rather make it "normal" (for you) to get better each year, or at least to not decline? Don't you want to feel the energy and focus you had at twenty-five when you're eighty?

Before I learned that it was possible to increase my brain energy,

I thought it was normal to get really pissed off in rush hour traffic, to wake up feeling exhausted after a full night's sleep, to get snippy with the people around me in the late afternoon, to crave sweets after a meal (isn't that what dessert is for?), to sometimes lose my train of thought midsentence, or to walk into a room and forget why I'd gone in there in the first place.

Maybe you only experience one or two of these symptoms on a regular basis. Most likely, they've become so "normal" to you that you don't even notice them until you start looking. You've figured out how to work around them so you can live your life—in fact, you're probably expending even more precious energy coming up with work-arounds so you can still function. But the truth is that none of these symptoms are normal. They are not inevitable. And they are not simply built-in mental weaknesses.

There is a way to change the amount of energy being delivered to your brain so that its energy level actually exceeds its demands. Once you learn how to do this, your brain can function like that brand-new phone, fresh out of the package—fast, responsive, and highly charged.

THE THREE F'S

Why do our brains require so much energy in the first place? The truth is, it's an evolutionary imperative—our brainpower is part of nature's plan to help us stay alive and propagate our species. If you were to design a species to live forever, it would need the built-in ability to do only three basic things, all of which are F-words: Fear things (deal with scary stuff in our environment using our "fight-or-flight" response), Feed (get energy from food), and the other F-word (reproduce!). Our bodies have evolved so that our species can survive just about anything the world throws at us, and our systems allocate energy to our cells the same way.

In the 1960s neuroscientist and psychiatrist Dr. Paul D. MacLean developed something called the "triune brain model," a simplified way of looking at the regions of the brain that is useful when we talk about how the brain uses energy. In this model, the "reptile brain" controls low-level processes like temperature regulation and electrical systems. Every creature with vertebrae has a reptile brain, and this part of the brain is first in line when it comes to energy needs. If you don't get enough energy and nutrients to this part of the brain, you will die, end of story.

All mammals share the second brain, which I refer to as our "Labrador retriever brain," because those big, happy dogs are such great examples of animals that bark at most things, eat nearly everything else, and try to mount what's left. Your Labrador brain controls the instincts that keep our species alive and propagating—the "three f's" that I mentioned earlier. Your Labrador brain means well. It is only trying to help you survive. The issue here is that the very urges that were meant to keep us alive can cause massive brain-energy problems.

You're probably familiar with the concept of "fight or flight"—our physiological response to a perceived threat. The ability to go into fight-or-flight mode was incredibly important when humans evolved, as lions and tigers were chasing us on a regular basis. Back then it would have been detrimental for us to stay focused on any single task when a pride of lions was lurking nearby. Our fight-or-flight response kept us a little bit distracted all of the time so that we could constantly scan the environment around us for threats. When our brains perceived a threat, they would divert all of our energy into the systems necessary to either kill a lion or at least run away from it faster than the slowest member of the tribe.

The problem is not only that lions don't pose much of a threat anymore but also that our bodies can't distinguish between real and perceived threats—they react the same way to any stimulus, from a lion to a bump in the night to an e-mail alert possibly delivering some bad news. And given our 24/7 lifestyles, we're now bombarded day and night with all kinds of stimuli—some completely harmless—that our biology compels us to respond to in the same way. This constant state of monitoring for danger and then overreacting to minor threats keeps the body in a constant state of emergency—sapping our energy, and therefore our focus.

A decrease in energy available to the brain triggers a brain emergency. After all, from the brain's perspective, if there's not enough fuel for the Labrador brain, then a tiger might eat you. So when energy in the brain dips, emergency stress hormones are released to steal energy from elsewhere in the body, and they make you feel like you want to either run away or kill something. You get distracted, yell at the people around you, forget what you were right in the middle of doing, and then give in to major sugar cravings—all things you'll be ashamed of after you have a snack.

When you resist the Labrador's urges, you are using the third and final part of the brain, your "human brain." This part of the brain—the prefrontal cortex—contains the most mitochondria, which is why all this resistance uses up massive amounts of energy. Every time you resist an urge, you are making a decision. Scientists have proven that there are a limited number of decisions you can make each day before you reach "decision fatigue."[2] Each decision requires energy, and when you're tired, hungry, or have already made a lot of decisions, you run out of energy and start making bad choices.

Being able to make good decisions is therefore a pretty good measure of brain performance. When you have enough brain energy, you'll be able to make better decisions for longer, and you'll be a lot less emotionally reactive when you don't want to be. Nothing will improve your life more dramatically than that.

A good portion of this book will show you how to turn down the Labrador brain so you can use your human brain to greater effect. Using the techniques I've outlined, you will learn how to make your cells more efficient at both creating and using energy. Even if you think your brain is already working pretty well, it will function a lot better when it becomes more energy efficient. Having stable brain energy helps turn off your Labrador brain because it stops sensing energy emergencies.

The overall goal is to grow a stronger and more resilient brain. This is a four-step process:

1. STOP DOING THE STUFF THAT MAKES YOU WEAK

Sure, this sounds obvious, but the problem is that most of us don't know exactly what we're doing to slow down our brains. Brain kryptonite is everywhere around us, from our breakfast to our bedside reading lamp. Overcoming the burden created by this brain kryptonite requires a tremendous amount of energy that your brain can't afford to lose. Identifying your personal kryptonite and removing it from your life will free up your brain's energy reserves for more important things.

2. ADD MORE ENERGY

Your mitochondria need oxygen to make energy, and they also need either glucose or fat (or sometimes amino acids). This does not mean that the more carbs you eat, the more energy your mitochondria will produce. Actually, the opposite is true. Your mitochondria perform best when they can alternate between fuel sources like a hybrid car. Through strategic dietary changes and supplementation, you can ensure that your mitochondria have the energy sources they need. Sorry to break it to you, but you're going to have to eat more creamy, delicious, satisfying fat.

3. INCREASE THE EFFICIENCY OF ENERGY PRODUCTION AND DELIVERY

Your mitochondria may not be producing energy as efficiently as they should be due to nutrient and antioxidant deficiencies, or as a result of damage from various toxins, stress, or even lack of sleep. There are specific things you can do to grow more mitochondria and make the ones you already have function better. Removing the toxins and brain kryptonite from your environment in step one will certainly help, and so will specific supplementation and changes to your diet and lifestyle.

4. STRENGTHEN YOUR MITOCHONDRIA

Mitochondria are typically referred to as the "powerhouses of the cell," and if you want to maximize your performance, you want your powerhouses to make as much energy as possible. (No one likes a weak powerhouse!) One of the most effective ways to do this is actually by stressing the mitochondria exactly the right amount. In the same way that lifting weights stresses and then strengthens a muscle, specific techniques place the right amount of strain on your mitochondria to kill off the ones that are past the point of no return and stimulate the remaining ones to grow stronger. I'll even share a few tricks that can help you grow more mitochondria than you have today!

THE FIVE BRAIN WEAKNESSES

Which of the steps above do you need to supercharge your brain? The answer is probably all of them, but you may need to focus more on one or two areas based on your particular issues. In my work with clients and my own experience of hacking my (admittedly weak) mitochondria, I've identified five main brain weaknesses that manifest in slightly different symptoms, depending on the individual. The one thing all of these symptoms have in common is that they are tied to mitochondrial function. You may have one or two or all of these weaknesses. Sometimes it's hard to tell, but it's important to know where you're starting from in order to get the most out of this program. Before I hacked my brain, I certainly suffered from all of these. The good news is, I don't anymore. That means you aren't stuck with the brain you have right now, either.

Maybe you're tired all the time. You still manage to get a lot done, but it requires an exhausting amount of effort just to keep up. But this probably feels normal to you, and you think it's that hard for other people, too. Or maybe you sense that things could be better, but you're not sure how bad it really is. What are the signs that your brain isn't functioning as well as it could?

The descriptions below will help you pinpoint the brain weaknesses that are impacting your performance the most. Identifying these weaknesses will allow you to understand exactly which areas of my program you'll need to focus on the most in order to become Head Strong.

BRAIN WEAKNESS #1: **Forgetfulness**

Do you pause or say "um" a lot when you're talking because you can't think of the right words? Do you open the fridge and then stare inside, wondering why you opened it in the first place, only to notice your car keys are in there? Or maybe you struggle with long-term memory—do you have a hard time placing people, recalling specifics about when or where something happened, or even remembering significant moments from your past?

Both short- and long-term memory loss stem from the same causes: not getting the right nutrition, chronic low-grade bacterial or fungal problems, inadequate neurotransmitters (the chemical messengers in your body), and, of course, impaired mitochondrial function. In this case, poor mitochondrial function can contribute to lower heart efficiency, which leads to low blood pressure or insufficient delivery of oxygen, fuel, and nutrients to the brain. Remember, your mitochondria need oxygen and food in order to produce energy. Ironically, this means that the worse your mitochondria are functioning, the less oxygen and food they'll receive. As a result, your mitochondria won't be able to produce enough energy to meet the brain's demands, creating brownouts in the brain, and your brain performance will suffer accordingly. This vicious cycle will go on and on until it begins to wreck your performance.

You'll experience this in small ways at first, like trying to think of a word but being unable to remember it. The effect is cumulative—I used to have dozens of brownouts a day. But once you fix your circulation and your blood pressure, your brain's energy production will improve. Your brain will get enough oxygen, your mitochondria will be able to make more energy, and you will be able to remember things much more easily. I never have brownouts anymore.

Even when your mitochondria problem is fixed, you'll still need healthy neurons, or nerve cells, to transmit messages in the brain quickly and efficiently. Brain-derived neurotrophic factor (BDNF) is a protein that supports the survival of existing neurons in the central nervous system and encourages the growth of new neurons and connections between them. Increasing BDNF through exercise, diet, and strategic supplementation will improve your learning, memory, and higher thinking.

Once you make those new neurons, you need the necessary building blocks to create myelin, the insulation around neurons that helps them send messages faster. Nerves without this insulation require much more biological energy than those with it. This is where nutrition comes into play. Your diet is a key factor when it comes to brain function. If you were building a high-end house, you would use quality materials, and your brain requires specific, nutrient-dense foods to create the highest-functioning connections in your brain.

So: is forgetfulness a problem for you? Ask yourself if any of the following symptoms feel familiar:

- Forgetting important dates or events on a regular basis
- Asking the same questions over and over
- Relying on memory aids more than you used to (checklists, electronic reminders, etc.)
- Difficulty keeping track of monthly bills
- Pausing while speaking to think of the right word
- Problems remembering names
- Trouble keeping track of regularly used items (phone, keys, etc.)
- Losing your train of thought often
- Forgetting what you are doing in real time—leaving a pot burning on the stove, picking up the phone and forgetting who you were going to call, and so on.

If this sounds like you, pay extra attention to the chapters ahead that focus on mitochondria function, oxygen delivery systems, myelination, neurogenesis, and nutrition.

BRAIN WEAKNESS #2: **Cravings**

When I talk about cravings, I'm not referring to the emotional cravings that stem from loneliness, boredom, or stress. I'm talking about a physiological urge that comes from your Labrador brain. These biological cravings are a sign that your brain needs energy.

Your mitochondria use oxygen to burn fat, glucose (sugar), or amino acids to make energy. If you eat too much sugar, your mitochondria will no longer easily produce energy from fat—they'll start producing all of your energy from glucose. This means that instead of your brain using fat as its fuel, the fat gets stored in your fat cells and you'll start to gain weight. Meanwhile, your brain will burn through glucose quickly so you get a blood sugar crash, which your inner Labrador interprets as an emergency, signaling the alarm for SUGAR and SUGAR, NOW. That is a craving in the making!

When I was fat, I was gaining weight and starving at the same time. The calories went into my mouth and straight to my fat cells instead of being used to make energy. I thought I was weak-willed, but I had fallen into a fat trap. Without adequate levels of energy being released from my fat cells, my mitochondria could not make enough energy, and my inner Labrador started begging me to eat more of everything.

I also didn't realize that toxins were a major cause of my food cravings. The kidneys and liver are your body's natural detox pathways. Any time you consume something that's toxic or to which your body is allergic, these organs send out an alarm asking for extra sugar to oxidize or metabolize (in other words, to neutralize and/or eliminate) the offending substance—and compete with the brain for glucose. This detox process leads to low blood sugar and results in cravings.

The same thing goes for any form of brain kryptonite. Very often, we sap our brain's energy without even realizing it. If you spend a lot of time in a noisy environment or one with bad lighting, your brain has to use a lot more energy to filter out all of the distractions. Remember, you can't store energy for more than a few seconds, so your brain needs a steady stream of glucose (or fat) to make it. When the demand goes up, your inner Labrador sends the signal, "I need sugar now!"

Have you ever taken your kid to a birthday party at Chuck E. Cheese's or gone to an amusement park and found yourself exhausted afterward and craving ice cream? That's because your brain had to work extra hard to filter out all of the stale air, background noise, and flashing lights (not to mention the toxic pizza and snack foods, if you dared to eat any). Your mitochondria probably aren't making energy as efficiently as possible, and thus they couldn't keep up with your brain's increased demand—so the Labrador panicked.

The more mitochondria you have and the more efficiently they function, the fewer cravings you'll experience. It's also essential to consume enough fat—and the right kinds of fat—so that the brain has multiple energy pathways and is not overly reliant on sugar.

So: are cravings a problem for you? Ask yourself if any of the following symptoms feel familiar:

- Frequent blood sugar dips throughout the day
- Strong desire to eat something sweet after a meal
- Inability to go more than two or three hours between meals
- Irritability when hungry
- Exhaustion after spending time in a noisy or chaotic environment

If this sounds like you, pay extra attention to the chapters ahead that focus on light, environmental toxins, and ketosis (the state your body is in when it is burning fat most efficiently to make energy).

BRAIN WEAKNESS #3: Inability to Focus

When you sit down to read or write something, do you find that you can only concentrate for a moment or two before you become distracted by thoughts, worries, or even things in the environment around you? When your brain can't focus the way you want it to, it's virtually impossible to perform at your peak. I suffered with this symptom for many years before I realized that it was largely a result of my fight-or-flight response being turned on when I didn't want it to be.

Our friend the Labrador doesn't care about the work you're trying to focus on or what your kids are saying to you. His job is to keep you alive, so he is busy sussing out potential threats in your environment. Is that blinking light on the stove the start of a fire? Was that "ding" signaling that you got a text message a sign of danger? Is that fly buzzing around your head an animal that's trying to eat you? (Labradors aren't that smart.)

Your more highly evolved human brain might know the difference between an approaching car and an approaching lion, but your Labrador doesn't, and he is always on alert, trying to keep you safe. What a good dog. The problem is, when your Labrador is always screaming "Emergency!" it's impossible to focus on the things you need to get done.

It gets worse when your brain doesn't get enough energy. Maybe you overindulged in a few too many beers and all of your oxygen- and nutrient-rich blood went to your liver to help process the alcohol. Now your brain feels like it's going to die because it has less energy. This is just as stressful to your brain as a tiger, and it signals yet another emergency.

When the brain is low on energy, it stimulates the release of cortisol (the stress hormone) and adrenaline (the fight-or-flight hormone) to make emergency fuel. The adrenaline breaks down muscle in order to access stored sugar reserves, which in turn signals your pancreas to release the insulin needed to metabolize that sugar. The resultant insulin spike creates an even greater brain emergency, and your brain triggers the release of more cortisol, and suddenly you want to *flee*! How could you possibly be expected to focus with all that going on?

Over time, this cycle can lead to a condition known as insulin resistance—a state in which your body becomes desensitized to insulin. When your body doesn't respond to insulin, your cells have trouble absorbing glucose, so it builds up in the bloodstream instead of being used to make energy. This creates unstable brain energy, which puts you in and out of fight-or-flight mode throughout the day. The result? You're easily distracted and can't seem to focus, no matter how hard you try.

It is possible to stop this cycle and end the fluctuations of adrenaline, cortisol, and insulin in your bloodstream by stabilizing your blood sugar throughout the day. In fact, during the writing of this book, I received test results showing perfect insulin sensitivity while on the Head Strong program—the very lowest score possible, 1 out of a possible 120. When you have stable blood sugar, it will help appease the Labrador, turn off your fight-or-flight response, and allow you to finally pay attention to what you want, when you want.

So: is inability to focus a problem for you? Ask yourself if any of the following symptoms feel familiar:

- Consistently interrupting others during a conversation
- Wandering thoughts when trying to concentrate
- Trouble completing tasks or meeting deadlines
- Being chronically late for appointments despite your best efforts
- Difficulty staying organized
- Inability to multitask efficiently
- Shifting from one topic to another in conversation

If this sounds like you, pay extra attention to the chapters ahead that focus on ketosis, meditation, and breathing exercises.

BRAIN WEAKNESS #4: **Low Energy**

Are you tired all of the time, or do you have a predictable energy slump at the same time every day? Do you find yourself moving slower than you'd like, as if you're trapped in quicksand? Or does your brain just feel foggy, like you're hungover or jet-lagged, even when you're not? These are all symptoms of the same brain weakness: low energy.

There are several things that may cause low energy, but the main culprit is poor blood sugar regulation. When you become insulin resistant and your body is unable to effectively process sugar, your brain pays the price. You'll have brain fog and fatigue and feel like life is passing you by. Luckily, it's relatively easy to regulate your blood sugar by following the high-fat diet in this program that teaches your body how to burn fat as fuel.

The other main cause of this brain weakness is inefficient mitochondria. No matter how stable your blood sugar is, if your mitochondria can't produce energy efficiently, you'll always be tired. Lucky for you, the Head Strong program is specifically designed to make your mitochondria more efficient at creating energy. If you follow my guidelines, you will notice an uptick in your energy level in as little as two weeks.

The final cause of low energy is brain kryptonite. Anything that pulls too much energy away from your brain and into another part of your body will leave you feeling sluggish. Changes to your environment that eliminate toxins will give you a needed energy boost.

So: is low energy a problem for you? Ask yourself if any of the following symptoms feel familiar:

- Afternoon energy slump
- Lack of mental clarity or "fuzzy" thoughts (brain fog)
- Fatigue and muscle weakness
- Sudden reduction in grip strength
- Sleep that isn't refreshing
- Extreme exhaustion after physical or mental exertion
- General malaise

If this sounds like you, pay extra attention to the chapters ahead that focus on ketosis, mitochondria, and environmental toxins.

BRAIN WEAKNESS #5: Moodiness/Anger

Most people don't realize that their mood swings and "uncontrollable" anger are a direct result of a weakness in their brain. Think about the triune brain model again. The most highly evolved "human brain" receives energy last, after the reptile and Labrador are both fed. As we discussed, this part of your brain—the prefrontal cortex—also requires the greatest amount of energy in order to function and therefore contains the highest density of mitochondria. Because of its massive energy requirements and the fact that it gets energy last, this is the part of the brain that usually suffers first when you don't have enough energy.

You've probably already figured out that it's the prefrontal cortex that helps to control your moods. This part of the brain is in charge of personality expression, decision making, and moderating social behavior. You obviously can't perform well if you're making bad decisions and acting poorly in social situations, so it's crucial for you to get enough energy to this part of the brain.

In retrospect, it's no surprise that my original SPECT scan showed almost no activity in my prefrontal cortex. Back then, my moods and emotions were all over the place. I had road rage and often snapped at the people around me. It didn't take much to set me off. Now that I've hacked my brain, my moods and emotions are completely different. I'm more patient, even-keeled, and full of joy.

Yes, joy. Even that is hackable.

So: is moodiness or anger a problem for you? Ask yourself if any of the following symptoms feel familiar:

- Highly active middle finger
- Tendency to snap at people over small things
- Lack of patience
- Depression
- Mood swings
- Quick temper
- Volatile behavior
- History of impulsive, poor decisions

If this sounds like you, pay extra attention to the chapters ahead that focus on mitochondria, environmental toxins, and brain kryptonite.

No matter where you're starting from, rest assured that the Head Strong program can help you. Just by using the techniques in this book, I went from suffering from all five brain weaknesses to having none of them. I can now count on my brain to work the way I need it to, no matter what is going on in my environment. The difference this has made on my life, career, and relationships is incalculable, and I can't wait for you to experience the same type of results.

Head Points: Don't Forget These Three Things

- Certain foods, products in our environment, types of light, and even forms of exercise can weaken your brain.
- All useless stimuli—potential threats, ringing phones, flashing lights, and so on—use energy in your brain.
- Forgetfulness, cravings, low energy, moodiness, and inability to focus are all symptoms of low brain energy.

Head Start: Do These Three Things Right Now

- Stop blaming yourself for running out of willpower—it's not a moral failing!
- Reduce the amount of stimuli in your environment when you want to focus—turn off the phone, limit alerts from your computer, cover the windows.
- Make the most important decisions first every day, before you can experience decision fatigue.

2

MIGHTY MITOCHONDRIA

One and a half billion years ago, the Earth was covered with warm seas and the air was filled with a terrible poison—oxygen—that killed most of the living organisms exposed to it. A few hardy species of bacteria, however, managed to adapt to these harsh conditions by learning how to use oxygen to make energy. These bacteria were able to take oxygen and create a substance known today as adenosine triphosphate, or ATP.

One of those species—which is believed to have been a small purple bacterium—eventually became embedded in another type of cell. Over the next billion years, those combined cells evolved to form animals and then humans. This ancient bacterium is still within us and continues to create ATP, the energy that our cells need in order to thrive. And new research shows that even today, these bacteria are in charge of what we do—to a greater extent than scientists ever expected. In fact, they call the shots on how you feel on a second-by-second basis.

What do we call these bacteria today?

Mitochondria.

In case you didn't have enough reasons to talk to your mother every Sunday, here's one more: she's the one who gave you all of your mitochondria. Many people think that we receive an equal 50 percent of our genetics from our mothers and 50 percent from our fathers, but we are actually more genetically similar to our mothers. When each of us was conceived, both the egg and the sperm contained mitochondria, but the mitochondria in the sperm—which powered its mighty swim toward the egg—got left behind when the sperm's tail dropped off as it burrowed into the egg. That means that the mitochondrial DNA (deoxyribonucleic acid) in the fertilized egg that became you came exclusively from your mom. When your yoga teacher talks about "divine female energy," neither of you probably realizes that she's actually referring to these ancient bacteria.[1]

Ancient bacteria. Divine female energy. Mitochondria seem pretty mysterious and magical, don't they? Let's take a moment to understand these tiny powerhouses.

Mitochondria are cigar-shaped parts of your cells, bound by a double membrane, with the inner membrane tucked and folded inside the outer membrane. The average human cell contains between one thousand and two thousand mitochondria. The cells in the parts of our body that require the most energy—the brain, retina, and heart—contain about ten thousand mitochondria each. That means you have more than *one quadrillion* mitochondria within your body. That's more than the number of bacteria living in your gut! And, in fact, our entire respiratory system—our heart, lungs, and blood—exists to deliver oxygen to our mitochondria so that they can make the energy (ATP) that keeps us alive.

Your mitochondria determine how your body reacts to the world around you. When your mitochondria become more efficient, your mental performance increases. The better your mitochondria are at creating energy, the better your body and mind will perform, the more you can do, and the better you will feel while doing it.

ATP—THE ENERGY OF LIFE

The most important thing your mitochondria do is extract energy from the food you eat, combine it with oxygen, and make ATP. We've only known about ATP for about a hundred years, and there's still a lot we need to learn. But we do know that ATP stores the energy required to power us both physically and mentally. Nearly all of your cells need ATP in order to function. Without it, they couldn't survive—and neither could you. The energy production that takes place within your mitochondria is therefore the single most important function in your body. ATP is your life's blood or, more accurately, the reason for your blood's life.

Think of it this way: You could live for at least three weeks without food. You could live for about three days without water. But without ATP, you would die within seconds.

The energy stored in ATP is released when it is used as fuel by the body. When your body uses ATP for fuel, it breaks down, creating two by-products: adenosine diphosphate (ADP) and phosphate (P). Remember, ATP is adenosine *tri*phosphate, meaning it contains *three* phosphate bonds. When two of these bonds are broken off into adenosine diphosphate and one lonely phosphate, energy is released. That energy is the power of you. And those quadrillions of little embedded bacteria actually control you.

After this process is complete, something amazing and elegant happens. Your body reattaches a phosphate molecule to the ADP, recreating ATP for it to be used as fuel and broken down again into ADP and P, releasing more energy. In essence, our mitochondria are the original molecular engines, using the same molecules to regenerate energy over and over. It's a much more efficient way of creating energy than making each unit of ATP from scratch.

If you're my father's age, this concept might sound familiar—it's strikingly similar to the process of a car engine turning over as it idles or accelerates. The first thing he did when he got a '57 Chevy was to figure out how to make it go a little faster. If you're my age, it doesn't sound so different from the methods my hacker friends and I used to overclock our gaming PCs so they would also go faster, while still war-dialing in the background.

But our mitochondria are a heck of a lot more powerful than a car engine or a computer processor. There are about one billion molecules of ATP in an average cell, and each molecule gets recycled about three times a minute. Even though you have roughly one hundred trillion cells, a normal person has only about 1.75 ounces (50 grams) of ATP in their entire body at any one time. Each mitochondrial ATP cycle can create about six hundred ATP molecules per second at maximum demand. That means that if you eat 2,500 calories a day, your mitochondria recycle and reuse those 1.75 ounces of ATP so many times that it's the equivalent of creating four hundred pounds of ATP over the course of the day.

In case creating energy for every system and function in your entire body wasn't a big enough job for your mitochondria, they're also in charge of other essential tasks such as transmitting signals between cells, cellular differentiation (the process by which one type of cell transforms into another), and maintaining the cycle of cell growth and cell death. When you think about it, mitochondria create all the power, control communication, and decide what lives and dies (and when). These little bacteria are actually calling the shots in your biology. I've started to see my own body as a big walking petri dish supporting a quadrillion mitochondria, doing whatever they want.

But your mitochondria have other skills, too. Mitochondria can change their shape and size, and some mitochondrial functions are unique to specific types of cells. For example, only the mitochondria in your liver contain an enzyme needed to detoxify ammonia, a waste product produced when the liver breaks down protein. The different parts of your body also use ATP from your mitochondria for their own specific functions. Your heart, for example, uses its energy to pump blood to your brain and the rest of your body, and your brain uses its energy to think, learn, remember things, and make decisions. Of course, the extra mitochondria in the brain require lots of oxygen to create ATP, so if the mitochondria in the heart are not producing energy efficiently, your brain will suffer from a lack of energy before the rest of your body does.

Cells like the ones in your brain, heart, and retina that are literally studded with mitochondria are the first at risk when you have less energy available than you need or when those cells waste the energy they were meant to use. When neurons have energy problems, you get cognitive impairment and brain fog. When cardiocytes (heart cells) have mitochondrial defects, you get heart dysfunction and feel tired. When myocytes (muscle cells) can't make energy, you see symptoms of fibromyalgia and chronic fatigue syndrome. When your enterocytes (intestinal cells) have energy problems, you see leaky gut and autoimmune diseases. And the list goes on. The critical systems in your body *all* rely on mitochondria to work. Or more accurately, your mitochondria control all of the critical systems in your body.

Are you convinced yet that you need to "mind your mitochondria," to quote my good friend Terry Wahls, a physician who wrote *Minding My Mitochondria*? She hacked her own mitochondria to reverse progressive multiple sclerosis. The exciting thing about mitochondria is that these essential structures are not in the least bit static. The mitochondria in every part of your body are constantly changing. They can be damaged, destroyed, improved, renewed, or outright hacked. There are many things you can do to make your existing mitochondria work better and even to literally grow more of these "power plants" in your cells.

My own mitochondrial function has been a focus of my biohacking for years, and I developed a wide array of habits to improve their function. In fact, every one of the biohacks I've found that has an immediate impact on my energy is a mitochondrial hack, including the ones that reversed my toxic mold exposure and chronic Lyme disease. If I'm starting to feel my concentration waver, I just implement one of my tools for improving mitochondrial performance, and soon I'm back in charge. In other words, when I want to kick more ass, I ramp up my mitochondrial function.

As I type this sentence, I'm on a stack of mitochondrial energizing supplements because it's almost midnight, I have four thousand more words to write before I go to sleep, and I'm recording two episodes of *Bulletproof Radio* tomorrow morning. My mitochondria *have to* kick ass to get all this done! I'll share my most important tools for improving mitochondrial performance in this book, including the ones I used while writing it.

GOOD MITOCHONDRIA GONE BAD

Before we learn how to upgrade our mitochondria, let's take a look at what causes mitochondrial dysfunction. After all, the easiest way to perform better is to just stop doing the things that slow you down.

The most predictable cause of mitochondrial function decline is aging. From age thirty to age seventy, the average mitochondrion decreases in efficiency by about 50 percent. That means that the average seventy-year-old is making about half the cellular energy as the average thirty-year-old. It's a good thing I have no intention of ever being an average seventy-year-old! This decline in mitochondrial efficiency contributes to almost every symptom and disease that makes getting older suck so much.

Perhaps that statistic slipped past you. A 50 *percent* decline in your energy level is considered "normal." But what if you maintained your mitochondrial performance so that it was the same at seventy as it was at thirty? You'd be the most ass-kicking seventy-year-old on the planet, that's what.

Here's the thing about mitochondrial decay: today it is considered to be inevitable. It's already started to slowly happen to you, with a speed dependent upon your genetics, your lifestyle, and the decisions you'll make about how to live your life from here on out. But the rate of that decay is not fixed. It is already theoretically possible to keep your mitochondrial efficiency stable well into old age, so that at age seventy you could be making the same amount of energy (or more) as you did at thirty.

The trick is to avoid early-onset mitochondrial dysfunction (EOMD) by supercharging your mitochondria now. EOMD was discovered and named by Frank Shallenberger, MD, one of the many lecturers I've learned from at the Silicon Valley Health Institute, the anti-aging nonprofit I've run for more than a decade. EOMD is defined as the deterioration of mitochondrial function in people under the age of forty. Dr. Shallenberger estimates that about 46 percent of people have EOMD.

One of the interesting things about EOMD is that most people who have it are asymptomatic. They are not yet suffering from any major symptoms, nor have they been diagnosed with any disease. They may have strong cravings, mood swings, and frequent exhaustion, but they don't feel sick. Over time, however, EOMD leads to accelerated cell death and cell loss, decreased cell hydration, increased free radical damage, decreased mental capacity, decreased ability of the body to detoxify itself, and mitochondrial decay—which means the mitochondria are destroyed. EOMD is reversible, but mitochondrial decay is not, so the earlier you can catch and reverse this condition, the better.

Here's what I want you to know: at *any* age, mitochondrial

dysfunction poses a real threat. It doesn't matter if you're under thirty or over fifty. If you want to enjoy an amazing life—not just a comfortable enough one—you'd better prioritize the health of your mitochondria like your life depends on it, because it does. Literally.

Early-onset mitochondrial dysfunction manifests itself in four main ways:

MITOCHONDRIA MISHAP #1: **Inefficient Coupling**

No, this isn't Gwyneth Paltrow's new name for divorce (though it is a pretty accurate term for my high school dating life). Fair warning—the material ahead is pretty geeky stuff, and you can skip ahead to Part II of the book if you're already convinced that your mitochondria are important and you just want to learn *what* to do to make them function more efficiently. But if you'll stay with me for a few pages, you'll learn *why* mitochondrial dysfunction happens in the first place and just how much power you really have over your own energy and your own brain.

The core process the cells in your body use to produce ATP is called the citric acid cycle or the Krebs cycle (named after the scientist Hans Krebs, who discovered it in 1937). The Krebs cycle is an incredibly complex, multistep process, but I will spare you the full flowchart because you don't need that level of detail in order to change your mitochondria. What you'll see here is a simplified version instead.

Before the Krebs cycle can begin, your body converts sugar (or sometimes protein) into glucose, or it converts fat into a ketone body (a water-soluble molecule that the liver produces from fatty acids) called beta-hydroxybutyrate (BHB). Both glucose and BHB can provide carbon and electrons, the raw materials that create energy. Those raw materials form a molecule called acetyl coenzyme A (CoA), and this is where the Krebs cycle starts.

Throughout each round of the Krebs cycle, your mitochondria oxidize CoA, creating carbon dioxide and electrons. These electrons "charge up" a molecule called NAD (nicotinamide adenine dinucleo-

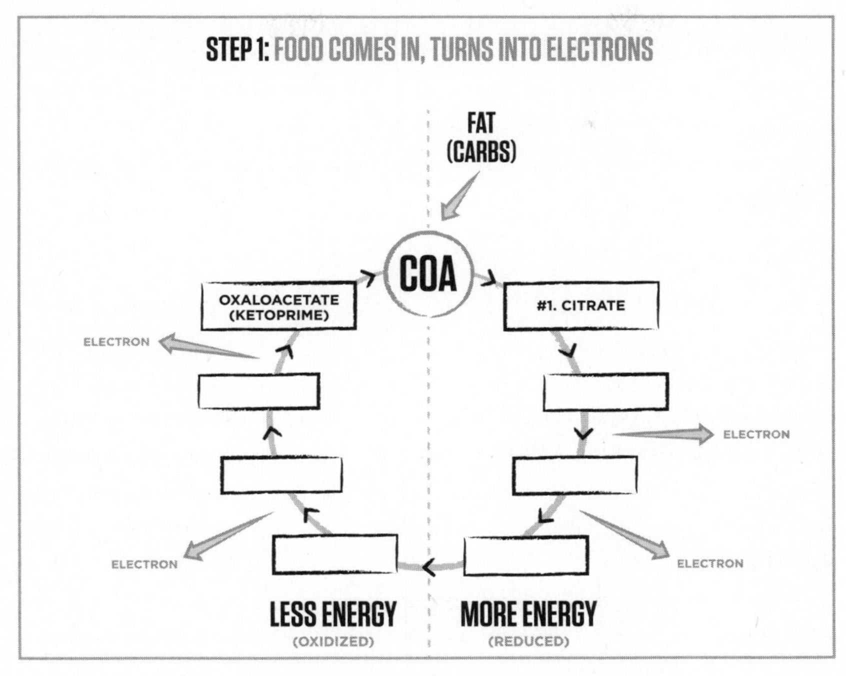

Food comes in, turns into electrons.

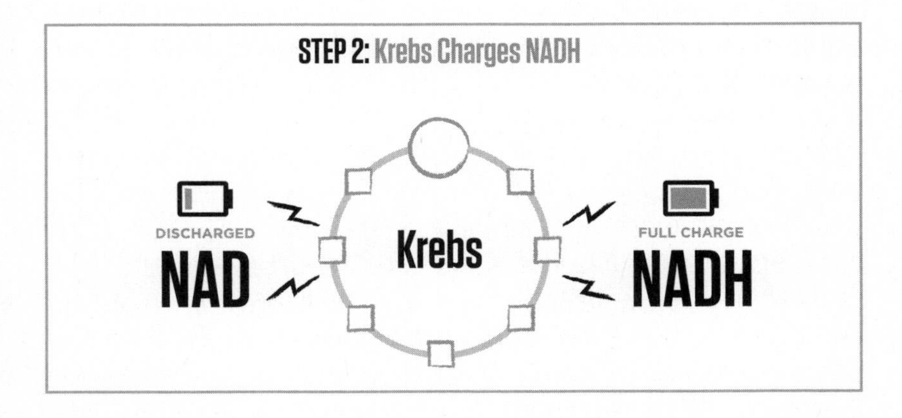

tide), which turns it into nicotinamide adenine dinucleotide, reduced (NADH). NADH is then one of the superstar molecules for your energy. Spoiler: there are "cheat codes" you can use to get more NADH!

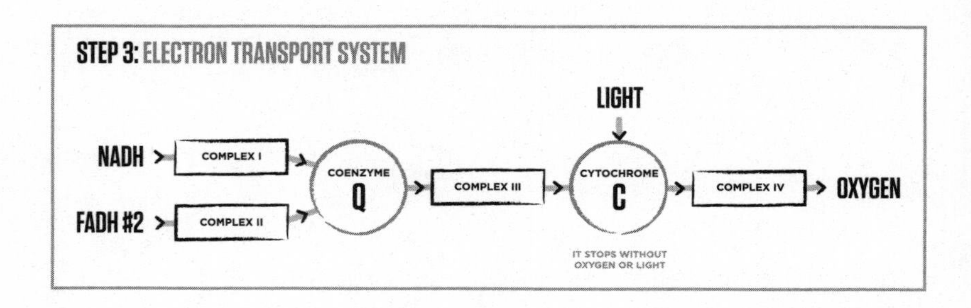

If you have lots of NADH, you'll like how you feel, because NADH is fully charged with electrons. It donates its electrons to the next step of the Rube Goldberg–like process that runs our biology, the electron transport system. There, molecules move electrons (negatively charged particles) and protons (positively charged particles) across the inner mitochondrial membrane, creating the power that drives the synthesis of ATP.

The protons and electrons must work in pairs—or as couples. Your body uses their attraction to each other as a source of power by putting a membrane barrier between them. If protons leak out, their partner electrons are left alone and useless—just like me in high school. Your body then uses oxygen to absorb those lonely electrons. But if the electrons and protons stay lined up on either side of your mitochondrial membrane awaiting their reunion, you don't need to waste oxygen absorbing the loners.

You can therefore measure your coupling efficiency by how much oxygen you use to create ATP. The more oxygen your body uses, the more protons are leaking, and the less efficiently your mitochondria are producing ATP. That makes you less efficient as well.

Even worse, using all of that oxygen to absorb single electrons creates free radicals that will damage mitochondria, slow you down, and give you a muffin top. A free radical, also known as a reactive oxygen species, is a molecule with a single, unpaired electron in its outer shell. These unpaired electrons make free radicals highly reactive to other substances and sometimes even to themselves. Because they are so reactive, free radicals can cause unwanted chemical reactions that damage cells. These reactions contribute to many diseases, including cancer, strokes, diabetes, Parkinson's, Alzheimer's, and schizophrenia. Free radicals are also a major cause of aging.

Inefficient coupling is one reason that type 2 diabetes increases the risk of heart disease. If you have type 2 diabetes, you have smaller mitochondria and fewer of them because they've been damaged by the free radicals that result from inefficient coupling. Remember, your heart is supposed to have a *lot* of mitochondria . . . but it won't have as many as it needs if you have diabetes.

When you are suffering from coupling inefficiency, your mitochondria will burn up a lot of oxygen to create ATP. This is unsustainable. Virtually all of the oxygen we breathe is used to produce energy in our cells by burning either fat or glucose. In the absence of enough oxygen to create ATP, your cells can produce energy anaerobically (without oxygen), but it is not as efficient. And, it can cause cancer—that's according to Otto Warburg, who won the Nobel Prize in Physiology or Medicine in 1931 for his hypothesis that cancer growth is caused by tumor cells generating energy anaerobically.

When your mitochondria don't have enough oxygen, they cannot recharge NAD by transforming it into NADH during the Krebs cycle, leaving an excess of NAD. When you have more NAD and less NADH, cellular aging is accelerated. The electron chain transport slows to a crawl, and you have more free radicals and less energy. Free radicals cause swelling in the cell, and the swelling makes the electron transport system less efficient. So you have even less ATP, and it is your brain that gets hit first as this vicious cycle of low energy takes hold.

The good news is that there are ways to prevent and reverse inefficient coupling. You are going to learn to make your mitochondria couple as efficiently as I wish I did in high school.

MITOCHONDRIA MISHAP #2: **Reduced Recycling**

Remember how your body brilliantly recycles ADP (spent ATP) by adding a phosphate molecule? Well, when your mitochondria aren't working perfectly, they use ATP more quickly than it can be recycled from ADP. Pretty soon, you end up with a buildup of ADP, creating a bottleneck in energy production. When this happens, the cell will run out of energy and need to rest until more ATP can be recycled from ADP.

Keep in mind that your cells cannot store more than a few seconds' worth of energy at a time; energy needs to be created on demand. Luckily, there is a Plan B for when your cells need energy and there's an ADP bottleneck in production. When this happens, your cells convert the available ADP into adenosine monophosphate (AMP). The problem with AMP is that it normally cannot be recycled. This is precisely why your body doesn't usually create it. You can think of AMP as disposable energy—inefficient and wasteful. Most of it is lost in your urine, and then you're back to square one with no energy and no ATP to make it with.

Your body must then create more ATP by recycling ADP or by creating it from scratch through the complicated Krebs cycle all over again. If things get really bad, your body can make a very small amount of ADP directly from sugar by converting it into lactic acid. One problem with this, though, is that it causes lactic acid to build up in the muscles, leading to pain and soreness. The other resulting problem is that it leaves no glucose available for the body to use. This means you don't have the raw materials to create new ATP. Converting glucose to lactic acid produces two molecules of ATP, but reversing this process to create glucose requires six molecules of ATP. It's the cellular equivalent of a farmer eating her seeds instead of saving them to plant the next season.

In short, inefficiently recycling mitochondria can create a complete metabolic disaster, and even small inefficiencies are going to show in your performance. Don't wait for Earth Day to make sure you're recycling! There is a brand-new hack for this recycling problem in the Head Strong plan.

MITOCHONDRIA MISHAP #3: **Excess Free Radical Production**

When your mitochondria are running like the amazing, high-performance semiconductors they are, they make ATP efficiently and create few free radicals. But when mitochondrial function is not as strong as it could be, mitochondria produce an excess of free radicals that leak into the surrounding cells and wreak havoc. This free radical damage lays the groundwork for many degenerative diseases.

Efficient mitochondria not only produce fewer free radicals, they also make special enzymes called antioxidant-buffering enzymes, which neutralize free radicals before they can do any harm. The problem is, these enzymes are made from ATP. When you experience a decrease in mitochondrial function, you end up with more free radicals and fewer of the enzymes needed to neutralize them. In simpler terms, your body is making too many bad guys and not enough of the good guys it needs to balance them out. It's a double whammy that ultimately causes mitochondrial decay, decreasing your energy production even more. Of course, there are several hacks for this, too!

MITOCHONDRIA MISHAP #4: **Poor Methylation**

Methylation is a mitochondrial process that occurs a billion times every second. (If you weren't already in awe of your body before reading this book, that's going to change!) During methylation, a single carbon and three hydrogen atoms (known as a methyl group) are added to another molecule. This relatively simple process controls your fight-or-flight response, sleep hormone levels, detoxification process, inflammatory response, genetic expression, neurotransmitters, immune response, and energy production. And it makes cell membranes, including the precious mitochondrial membrane that holds your electron transport system.

The methylation process also creates amino acids—critical elements of cellular energy production—as well as the ADP that your body converts to ATP. If your methylation process is impaired, so is your body's energy production. To make matters worse, your body needs ATP in order to methylate. This is yet another vicious cycle: you need ATP to methylate, and you need to methylate to make ATP.

During methylation, your body creates an amino acid called carnitine that is essential for breaking down fatty acids to use as an energy source. When you have poor methylation and your ability to create energy from fat declines, you start producing most of your energy from glucose. Because you're not burning fat efficiently, your body starts storing it, causing you to gain weight. And because you're making all of your energy from glucose, your blood sugar becomes less stable, you eventually become insulin resistant, and your inner Labrador panics, begging you to eat more sugar. Many people find that they start accumulating fat more quickly around middle age. That's partly because their mitochondria aren't burning fat efficiently.

CAUSES OF MITOCHONDRIAL DYSFUNCTION

Each of these mishaps is disastrous for your body's energy production and performance, but the good news is there are hacks to reverse all of them. Let's take a look at the leading causes of EOMD and some of the ways we can fix them . . . or prevent them from occurring in the first place.

NUTRITIONAL DEFICIENCIES

You need to feed your mitochondria with the highest-quality raw materials possible so that they can make energy efficiently and rebuild themselves when they are damaged. Mitochondria require many different nutrients, which I'll discuss in detail later, but for now you should know that proper nutrient intake is one of the easiest and fastest ways you can boost your mitochondrial function. When you flood your body with the right nutrition, your mitochondria might just perk up and come back to life—and so will you. As long as there isn't something else in the way.

HORMONAL DEFICIENCIES

Mercury poisoning, liver problems, and fluoride can all lower your thyroid hormones, which are essential for maintaining mitochondrial function and efficiency. Your liver converts T4, the main thyroid hormone, to T3, which helps your mitochondria create ATP. If your liver isn't functioning well, it won't create enough T3 to make energy efficiently. When I weighed three hundred pounds, I had extremely low thyroid hormone levels. If your energy is really low, you owe it to yourself to get an advanced thyroid test. And for some of the best information about thyroid and energy you can find, check out the book *Hashimoto's Thyroiditis: Lifestyle Interventions for Finding and Treating the Root Cause* by Izabella Wentz, the Thyroid Pharmacist.

Another hormone that impacts mitochondria is insulin. When your blood sugar is consistently high, your pancreas pumps out more and more insulin to try to control it, but you ultimately stop using that insulin efficiently and become insulin resistant. Insulin fluctuations signal your body to release the stress hormone cortisol, which inhibits fat metabolism. Your mitochondria start to burn sugar exclusively to create ATP, which is a less efficient energy source than fat. And as your body requires more and more sugar to meet your energy demands, you get blood sugar swings, causing your inner Labrador to panic. You give in and binge on doughnuts—which of course makes everything so much worse.

Because your mitochondria make ATP more efficiently from fat than they do from sugar, fatty acids are an important source of fuel. Fatty acids are stored in your adipose tissue (the anatomical term for fat) as triglycerides (fats found in the blood). Between meals, your body breaks down triglycerides into glycerol and free fatty acids, generating acetyl CoA, which, you'll remember, is the entry molecule for the all-important Krebs cycle. This means that if your body cannot efficiently metabolize fatty acids and break down fats, it does not have access to the ideal raw materials to create ATP.

Besides your diet, what determines whether or not your body can efficiently break down fat and use it for fuel? Your hormones. Certain hormones speed up your ability to break down fat, while others slow it down. Most people are unaware of the fact that your mitochondria are responsible for all of your steroid hormones, such as testosterone and estrogen. But, in fact, the inner mitochondrial membrane converts cholesterol to pregnenolone, the "mother hormone" that is the precursor to all the steroid hormones made in your body.

Improving mitochondrial function increases testosterone, and taking testosterone reduces oxidative stress in the brain,[2] which is a sign that your mitochondria are working better. That makes sense, given that your mitochondria have receptors for estrogen, testosterone, and thyroid hormones. The number of mitochondria you have in some cells is dependent on your testosterone levels[3] and growing new mitochondria can come from estrogen.[4] But as you age, your mitochondria make less testosterone . . . which means you make fewer mitochondria . . . which means you make less testosterone. Yikes.

A study done in 2013 on monkeys supports the ideas of the Head Strong plan. It showed that dysfunctional, misshapen mitochondria in the front of monkeys' brains harmed cognitive performance. And hormone replacement improved mitochondria and at the same time improved cognitive function in those same monkeys.[5] That's not to say that you must have hormone replacement or bioidentical hormone therapy to be Head Strong. In fact, when you hack your mitochondria, your sex hormones may go up enough that you don't need to go on hormones. But going on bioidentical hormones (testosterone and thyroid) did indeed help me turn my brain on back when I was twenty-seven. At the time, I didn't know why it had such a profound effect on my brain. Now I do—I was hacking my mitochondria.

Thyroid problems are rampant right now, and they're affecting our mitochondria. I had Hashimoto's thyroiditis with very low thyroid levels, which is well known to make you feel exhausted. I still remember the first day I tried thyroid hormone. I felt like I got my brain back that day. I didn't realize at the time that thyroid hormones cause a rapid stimulation of mitochondrial function right after you take them, and then after you take them for a few days your body starts to grow new mitochondria, and the mitochondria you have get bigger.[6]

The cool thing about your mitochondria is that when you make them more powerful, you get better at *everything* you do, including making hormones. While the cells in the brain, retina, and heart have the most mitochondria in men—ten thousand each—women have ten times that in their ovarian cells. That's right—some cells in the ovaries have one hundred thousand mitochondria each.[7] This may further explain why mitochondrial enhancement is so powerful for women and why poor mitochondrial function is associated with hormonal irregularities.

The hacks in this book will help you optimize your hormone levels and keep your blood sugar stable. You'll be surprised by how much this impacts the way your brain performs on a daily basis.

TOXINS

Environmental toxins are a leading cause of mitochondrial dysfunction. We are now exposed to thousands of toxic chemicals and pollutants that simply didn't exist a mere hundred years ago. These chemicals have made their way into our bodies, and our mitochondria have not evolved to thrive among them. If you want to kick more ass, you must eliminate these pollutants. Toxins that hurt mitochondrial respiration even a little bit have no place in our homes, in our food, in our coffee, or in our lives.

Your body requires a lot of energy to detoxify and expel or neutralize these toxins. Anything you can do to increase your cellular energy production, therefore, can also enhance your body's ability to detox. But given today's influx of toxins, you can't get by with the same amount of cellular energy that would have been serviceable a hundred years ago. Dr. Frank Shallenberger (who discovered and named EOMD) estimates that we actually need 50–100 percent more energy now than we did a hundred years ago to get rid of all of the toxins that are inside our bodies, slowing down our energy production and making us weak.

Heavy metals such as lead and mercury are some of the top offenders. Unbeknownst to me, I was suffering from environmental mercury and lead poisoning when I was at my weakest. Removing those heavy metals from my body helped me feel more energized. Heavy metals are stored in fat, and thankfully there are ways to mobilize fat and get rid of them. You have to do this carefully, though, so that the toxins aren't released inside the body instead of being expelled from it.

Our bodies actually create toxins that are just as damaging to our mitochondria as the chemicals in our environment. As with most types of fuel, the energy production process in our bodies creates some dangerous by-products—and mitochondria cleverly produce antioxidants and other detoxifying enzymes to counteract these by-products.

This is a delicate balance. When you don't have enough antioxidants present to counteract the free radicals in your body, you begin to suffer from oxidative stress. Oxidative stress is a sign of mitochondrial problems, and scientists believe it is a cause of many diseases including cancer, ADHD (attention deficit/hyperactivity disorder), autism, Parkinson's, Alzheimer's, chronic fatigue syndrome, and depression. Glutathione is a protective antioxidant that serves as mitochondria's main line of defense against damage from oxidative stress, but sometimes your body doesn't make enough of it. There are, however, ways to get your mitochondria to increase their production of antioxidants like glutathione, and you can supplement with it—I do both.

Your body also has a built-in detox process to recycle damaged cellular components. This is called autophagy, a Greek word that translates as "self-eating." During autophagy, your cells scan the body for pieces of dead, diseased, or worn-out cells, remove any useful components from these old cells, and then use the remaining molecules to either make energy or create parts for new cells. This janitorial process removes unwanted toxins, lowers inflammation, and helps to slow down the aging process.

Mitophagy is one stage of the autophagy process. This is the selective degradation of mitochondria. You may assume that you want to hold on to all of the mitochondria you possibly can, but it's actually much better for your cells to get rid of the ones that aren't working that well. You can think of the mitophagy process as similar to deleting old pictures from your phone. It will work better and faster if there is less junk clogging up the system.

I have focused a lot of my work on hacking autophagy, and I can feel the difference in my energy when my cells detox themselves efficiently. Boosting your autophagy process is one of the most important things you can do to improve your performance—which means it's definitely part of the Head Strong program!

STRESS

Real or perceived, physical or psychological, any type of stress causes your adrenal glands to release cortisol, a hormone that helps control your blood sugar levels, metabolism, immune response, inflammation, blood pressure, and central nervous system activation.

Cortisol isn't inherently bad. We need a certain level of it at all times, and we need more of it during times of stress—that's why your body cranks up the cortisol as part of your fight-or-flight response. Once things calm down, your cortisol levels should go back to normal. The problem is that for many of us, things never calm down. Our stress response is activated so frequently that our cortisol levels remain high all of the time. Chronic stress leads to many problems, including poor fat metabolism and an increased demand for sugar. Lowering your stress levels will help to quiet your inner Labrador and improve your ability to metabolize fat. And, given that your mitochondria feel and respond to your stress, lowering your stress allows your body to use energy more efficiently.

There are, however, certain types and amounts of stress that are beneficial to your mitochondria. Autophagy happens in response to mild stress from exercise or calorie restriction. Cellular stress also activates mitochondrial biogenesis, the creation of new mitochondria. That's right—you are not forever stuck with your old, lame, dysfunctional mitochondria. By strategically and temporarily stressing your body, you can boost its natural detoxification systems and stimulate the creation of new mitochondria to maximize your energy. Even better, there is a hack for calorie restriction so that you can reap the benefits without ever feeling hungry.

It is essential, however, that you don't stress your cells too much. When a cell is taxed, it begins a suicidal process called apoptosis, or programmed death. At the start of this process, the mitochondria release proteins that effectively schedule the death of the cell. Once these proteins are released, this process is irreversible. Yet, apoptosis isn't always perfect. Some cells linger beyond their "use by" date instead of submitting to apoptosis, and these cells continue to replicate, often becoming cancerous or diseased. Other cells die before they are scheduled to. Apoptosis malfunction is connected to cancer, autoimmune diseases, Alzheimer's, Parkinson's, inflammation, and viral infections.

Your mitochondria hold the signaling proteins that induce apoptosis, so anything you do to boost your mitochondrial function will help you hold on to healthy cells and get rid of the ones that are making you weak. Eating specific foods will also help to induce apoptosis in unhealthy or diseased cells.

I designed the Head Strong program to help you eliminate the causes of mitochondrial dysfunction, boost your mitochondrial efficiency, and help you grow new mitochondria through the following pathways:

- The best possible nutrition
- Oxygen therapy through proper exercise and boosted circulation
- Stabilized blood sugar levels
- Optimization of hormone levels
- Effective detox and avoidance of toxins
- Functional amounts of stress that induce autophagy and apoptosis
- Exposure to better-quality light and elimination of poor-quality light
- Changes to the water in your body

When you make these relatively simple changes, you will feel an incredible difference in your mood, your energy, and your overall performance. As a side effect, you might look better, too. This is the power of your mighty mitochondria.

Head Points: Don't Forget These Three Things

- The cells in your brain, heart, and retina have the most mitochondria and are the ones to suffer first when energy demand exceeds supply.

- Your hormones, blood sugar levels, diet, and lifestyle all affect the function of your mitochondria.
- From age thirty to age seventy, the average person experiences a 50 percent decline in mitochondrial efficiency.

Head Start: Do These Three Things Right Now

- If you have big problems with energy, have your advanced thyroid hormone levels checked by a functional medicine doctor.
- If your energy crashes after meals, check your blood sugar levels either using a home glucose meter or a functional medicine doctor.
- Pay attention to your energy dips throughout the day—perhaps you ate something or were exposed to something that damaged your mitochondria!

3

BECOME A NEUROMASTER

Own Your Neurons

You probably learned about neurons back in high school biology class and haven't thought about them since. At the time, your teacher may have described the primary job of neurons as connecting to other neurons to create what's known as "neural networks." But I'm guessing you weren't taught that the way your neurons function and connect to one another determines how quickly you think, respond, and even learn. Or that your neurons play a huge role in your daily performance. Or that you are at least partially in control of that process. Because, as it turns out, there's a lot you can do to change how your neurons work right now and in the long term.

Why are neurons such a hackable part of your performance? Two reasons:[1]

First, your neurons are energy-sucking miracles of cellular engineering. One neuron in your brain uses up to 4.7 billion ATP molecules per second.[2] When scientists isolate neurons in a lab and give them an insufficient amount of ATP, their functions become unpredictable.[3] In fact, neurons can die if they don't have a constant supply of ATP,[4] as everything they do requires massive amounts of energy. So it would follow that if you can increase the amount of ATP produced by your mitochondria, you can enhance the performance of your neurons. Because who wants an unpredictable performance?

Second, each of your neurons is made up of a tiny cell body with little branches that extend from it (more on that in a minute). Those microscopic branches can extend up to three feet from a single cell! Not only do neurons do incredibly energy-intensive tasks, they have to do them across a long distance. Neurons contain two different kinds of motors designed to move mitochondria around inside the cell, and these motors also require energy.[5] Up to 30 percent of the mitochondria in your neurons are moving around to deliver their energy,[6] like backup generators sent on trucks to meet the high demands of an electrical grid and prevent a brownout. And studies have shown that there is a link between the slowing of these motors and the likelihood of developing neurodegenerative diseases.[7]

Like all other cells in your body, a membrane made of tiny fat droplets surrounds your neurons. But structurally, your neurons are unlike any other human cell. Given that one of their jobs is to send and receive messages to other cells, neurons have unique cellular components called dendrites and axons. Dendrites branch out from each neuron to receive information from other cells. They are the "ears" of the neuron because they listen to messages from all over the body. Axons, on the other hand, send information to other neurons—they're the "voice" of the cell because they do the talking.

But that information doesn't just plug in directly from one neuron's axon to another's dendrite. There are gaps between neurons called synapses that transmit messages from one cell to another. Synapses rely on chemical messengers called neurotransmitters and—you guessed it—a lot of mitochondria to fuel the process.

This cycle of sending a message from one neuron's axon into a synapse and then to another neuron's dendrites is the basis for how your brain functions. It is primarily a chemical and electrical process. When a neuron in your brain is at rest, the inside of it contains a negative charge and the outside has a positive charge. The cell membrane keeps these positive and negative charges separated by selectively allowing ions (charged atoms or molecules such as calcium, sodium, chloride, and potassium) to move in and out of the cell. The negatively charged ions inside the cell are not allowed out, while positively charged ions can move freely back and forth across the cell membrane. This balance keeps the cell negatively charged except when it's time for some action.

When the neuron wants to send a message to another neuron, its cell membrane allows positively charged ions to flood into the neuron, which changes it from negative to positive and makes it "fire," causing an electrical signal to shoot down the axon. After your neuron fires, its membrane gets back to work reestablishing its negative charge by pumping positive ions out of the cell while keeping the negative ones inside. Once the neuron is safely recharged to its negative state, it has the potential to fire again.

Interestingly, neurons cannot fire a little bit—every time one of your neurons fires, it does so at its full potential. It does this to ensure that the signal it's sending doesn't weaken as it travels through the axon and across the synapse, hopefully arriving at the dendrites of another neuron with enough strength to be heard. At least, it works that way if your axons are sufficiently protected by a fatty coating called a myelin sheath, which insulates the signal. (We'll look at what you can do to generate more myelin in just a bit.) And, of course, there must be sufficient ATP to power the whole process.

Meanwhile, the electrical impulse sent through the neuron's axon stimulates neurotransmitters to flow into the synapse. These neurotransmitters help the listening dendrites of the next neuron receive the message and fire in turn to relay it to the next neuron. This is how your brain communicates a message: through a path of connected neurons, one neuron at a time.

Together, many of these neural connections or pathways form vast neural networks. They are the seat of all learning and memory, carrying information from your short-term memory to your brain's structural core, where they are stored as long-term memories. If you have difficulty learning new things, remembering everyday tasks, or recall-

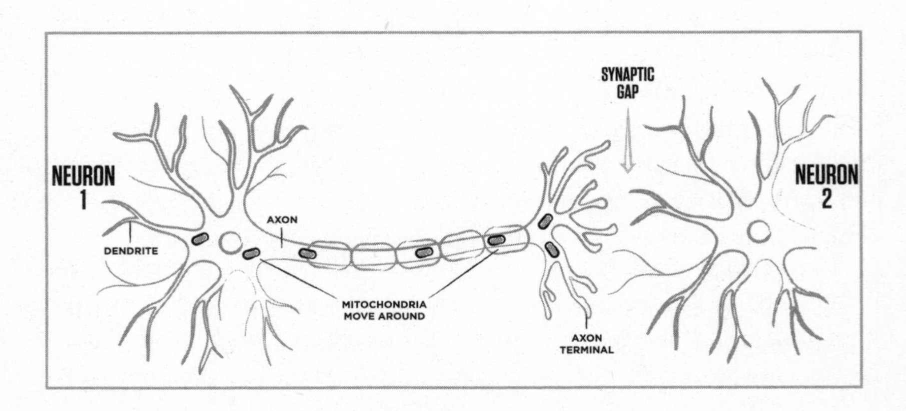

ing memories from the past, the strength and functionality of your neural networks may be to blame.

These networks are not only critical to your ability to learn new things and remember others; they are also critical to your ability to focus. In fact, a McGill University researcher recently discovered a network of neurons in the prefrontal cortex—your human brain—that is responsible for filtering out visual information and other distractions.[8] If this group of neurons isn't performing efficiently, your Labrador brain remains in a constant state of high alert, ready to respond to each piece of unfiltered stimulation as if it were a threat to your very existence. Of course, being distracted by all of those excess stimuli makes it pretty tough to concentrate on getting your job done.

Thankfully, there are a lot of things you can do to improve the function of your neurons, including helping to build those protective myelin sheaths that insulate the communication pathways between neurons, creating brand-new, healthy neurons, or just providing more power to your neurons.

HOW YOUR NEURONS CAN PERFORM BETTER

As you know, all cells have a membrane made primarily of fat, but myelin is a special, thicker fat layer that is essential for your brain to function; without it, the signals between neurons would simply be lost.

We are born with very little myelin, and the process of producing it (called myelination or myelinogenesis) occurs rapidly during infancy. This is the main reason babies develop so quickly from adorable little lumps to walking and talking humans. On the opposite end of the spectrum, demyelination, or the loss of myelin along the cells' axons, is responsible for many neurodegenerative diseases, such as multiple sclerosis.

Researchers have learned a lot about how myelin is created (and destroyed) by studying patients with MS. While the majority of us are unlikely to develop MS, we can all benefit from this research by looking at the methods doctors are using to help patients with neurodegeneration. It takes a lot more work to restore failed myelin than it does to keep yours strong, but the methodology is the same.

A type of brain cell called oligodendroglia (say that three times fast) does the work to form your myelin. Throughout adulthood these cells constantly generate new myelin and replace segments of myelin that break.[9] Just as the electric company maintains power lines to ensure smooth signals across a network, so, too, do these specialized cells conduct myelin maintenance throughout your neural networks. And your maintenance worker cells rely on a proper balance of hormones—particularly thyroid hormones and progesterone—to do their job well.

In chapter 2 we discussed how thyroid hormones affect mitochondrial function and ATP production. Research shows that thyroid hormones are also crucial for the health and survival of oligodendroglia cells and therefore for the constant rebuilding of healthy myelin.[10] This is one more reason to stay on top of your thyroid health and get your levels checked once a year or more often if you are feeling tired a lot.

Researchers have found that the hormone progesterone—which we typically associate with female body chemistry, since it's involved in the regulation of menstrual cycles—signals oligodendroglia to initiate the process of remyelinating neurons. In one study, mice that were treated with progesterone had more oligodendroglia cells that were able to repair more myelin.[11]

Progesterone is not limited to the female body—it is also present in males and is actually a prerequisite for sufficient testosterone production. If you have low progesterone as a man, your hair falls out, you get fat, and you grow man-boobs. (Trust me on that last one, as a former three-hundred-pound obese guy with hormones that were out of whack.) A functional medicine practitioner can run a hormone panel to check your progesterone level. I recommend that men and women over thirty-five get tested if they're having symptoms of low progesterone, such as brain fog.

In addition to the right hormones, your oligodendroglia cells also need the right raw materials to rebuild myelin. Remember, myelin is made of fat—specifically saturated fat, cholesterol, omega-3 fatty acids, and a few omega-6 fatty acids. This is one reason that Western medicine's war on cholesterol has been so harmful and misguided. Cholesterol is *essential* for cognitive function. Your brain makes up only 2 percent of your body weight but contains 25 percent of your body's cholesterol.[12] Most of that cholesterol is in your myelin, which is one-fifth cholesterol by weight.

Eating enough fats—and the right kinds of fats—is crucial to maintaining your myelin and keeping your brain's signaling both fast and efficient. Cholesterol deficiency is linked to a decline in cognitive function and memory, especially for anyone following a high-carbohydrate, low-fat, and low-cholesterol diet (also known as the standard American diet—which truly is SAD). A study in the *European Journal of Internal Medicine* showed that a diet deficient in dietary fats and cholesterol and high in carbohydrates could contribute to the development of Alzheimer's disease.[13] On the other hand, eating a diet that's high in the right kind of fats has been shown to increase memory and cognition[14] in healthy patients and even in some patients with signs of neurodegeneration.

I recently had the pleasure of interviewing Dr. Terry Wahls. Her story shows exactly how important a high-fat diet is for myelin production. In the year 2000, Terry, a practicing physician, was diagnosed with multiple sclerosis. Remember, the primary symptom of this disease is myelin degeneration. Terry sought out the best medical care, underwent chemotherapy, and took all the recommended prescription medications. Yet by 2003, Terry's disease had progressed to the point where she had become disabled and she relied on a wheelchair. Her doctors were very clear with her that once her faculties were lost, they couldn't be regained, and she herself had learned the same thing in medical school. She assumed that she would simply become increasingly disabled until she was bedridden.

In 2007, Terry became frustrated with the lack of options she was getting from her doctors and began researching her own disease as well as the brain in general. She learned everything she could about the fat composition of myelin and the exact nutrients that the oligodendroglia cells need to maintain healthy myelin. Terry put together a food plan for herself that included lots of healthy saturated, omega-3, and omega-6 fats and vegetables, and a bunch of other therapies. Within months, she went from being wheelchair-bound to riding a bicycle eighteen miles. The next year, she completed a trail ride in the Canadian Rockies. Today she is fully recovered. She got rid of her wheelchair; she can walk easily, and even jog. Score one for fats and for Terry, who now uses her protocol to help other patients suffering from neurodegenerative diseases.

There is also evidence that the right kind of temporary stress, particularly a diet that mimics the effects of fasting, can improve myelination. In a new study out of the University of Southern California, researchers looked at both mice and humans with multiple sclerosis to study the impact of fasting on neurons. They found that a fasting-mimicking diet promoted myelin regeneration.[15] In mice, the diet triggered autophagy, killing the bad cells that were causing damage to the myelin in the first place and stimulating the creation of new myelin. All of the mice showed an improvement in symptoms, while 20 percent reached a fully disease-free state.

Interestingly, when the fasting-mimicking diet was tested on humans, the control group was placed on a high-fat diet. In this case, both groups experienced improvements in their mental health. If a high-fat diet and a fasting-mimicking diet can both separately help people with major myelin decline regenerate their myelin, imagine what combining those two techniques can do for your brain. This is exactly what you'll experience on the Head Strong program.

There is one more way to improve your body's myelin, though, and it is through an unlikely source: your gut bacteria.

Remember the old saying that the way to a man's heart is through his stomach? Well, I say that the way to a man's (or woman's) *brain* is through his or her gut. Over the past ten years, we've benefited from exciting breakthroughs in our understanding of the bacteria that live in our intestinal tracts and the many vital roles they play throughout the body. Research has revealed that the gut and the brain communicate with each other constantly. In fact, gut microbes play a significant role in the way the brain functions and even the way it is wired.

In a brand-new study,[16] researchers analyzed how gut microbes affect gene activity in certain parts of the brain, particularly in the all-important prefrontal cortex. They compared gene expression levels in germ-free mice to those in normal animals and found that approximately ninety genes expressed themselves differently in the germ-free animals. To their surprise, many of these genes were known to be involved in myelination and were more active in the animals' prefrontal cortex. When the researchers dissected the animals' brains, they saw that the neurons in the prefrontal cortexes of the germ-free mice had thicker myelin sheaths than the normal mice.

In other words, there is a direct relationship between the bacteria in your gut and the myelin in your brain's prefrontal cortex, and it appears that some gut bacteria actually inhibit neuron function—and, therefore, your mental abilities.

Researchers are still studying exactly what in the microbiome is driving this difference, but the implications of this study are incredibly exciting for a biohacker like me. Perhaps in a few years we'll know exactly which probiotics we need to take in order to grow new myelin. In the meantime, it's clear that eating a diet that's high in fat keeps your hormones balanced, mimics the effects of fasting, fuels mitochondria, and supports healthy gut bacteria that will hot-wire your brain for maximum performance.

BUILDING NEW NEURONS

Maintaining healthy myelin will enhance the performance of the neurons you already have, but it's also possible to grow new neurons. The creation of new brain cells is called neurogenesis. Until the late 1990s, when I was still working at the company that held Google's very first server, scientists believed that neurogenesis ended in our late teens or early twenties. But we now know that the brain can produce new cells throughout our lifetime, just like Google can add more servers to their network. Brain degeneration is not inevitable. Brand-new, healthy neurons can be created at any age. It just takes more conscious planning to keep doing it as you age.

Your rate of neurogenesis, or how frequently you make new brain cells, is an important marker of brain performance. A low rate of neurogenesis is associated with cognitive decline, memory problems, and even anxiety and depression. A high rate of neurogenesis, on the other hand, is associated with just the opposite—cognitive enhancement, rapid learning, quick problem solving, and robust emotional resilience against stress, anxiety, and depression. That's what we all want, right?

Well, we're in luck, because it is possible to significantly increase the brain's rate of neurogenesis. When I interviewed Brant Cortright, PhD, a bestselling author and expert on neurogenesis, he told me that anyone can increase their rate of neurogenesis by at least five times.

Mic drop.

Did you catch that? Your brain can be *five times* better than it currently is at replacing defective cells or growing new ones. It's okay to set down the book for a minute to high-five someone you don't even know. That's what I did when I realized how much untapped potential there was in my brain!

So how do we reach that potential? First let's look at how new neurons are made. This is cutting-edge stuff because just twenty years ago, no one even knew that neurogenesis was possible. Then, in 1999, scientists discovered that neurogenesis was taking place in the brains of monkeys. Monkeys—and humans—form new neurons primarily in the hippocampi, the lateral ridges on either side of the brain. One end of the hippocampi helps to regulate emotions, especially stress and depression. The other end governs thinking and cognition. The hippocampi are also rich in neural stem cells, which give birth to new neurons. (As part of the research for this book, I had my own stem cells injected into my brain, but that's a story for another chapter.)

To create new neurons, neural stem cells divide in two. This division either produces two new stem cells, two early progenitor cells that will later differentiate into another type of cell, or one of each. When a stem cell divides and creates another stem cell, that new stem cell can continue dividing and creating more and more stem cells. If the stem cell divides and creates an early progenitor cell, that new cell then becomes a specialized cell—a cell with a specific job to do, such as an oligodendrocyte cell. Since oligodendrocyte cells are in charge of myelin production and maintenance, this of course means that your rate of neurogenesis has a direct impact on the integrity of your myelin.[17] In other words, when you make more neurons, you automatically keep more of your precious nervous system insulation intact without having to do any more work. It's a free upgrade!

Once a new cell is differentiated, it must settle into the existing neural network, much like the new guy or girl at work who has to learn the office culture and figure out how to fit in. The neuron develops an axon and dendrites and begins to communicate with other neurons. It can take four to six weeks for new brain cells to mature and integrate into the circuitry. This is like a probationary period for your brain's new hires. As they are becoming differentiated and learning how to do their jobs, new neurons run a high risk of premature death. This is not autophagy, which protects you by killing off and recycling unhealthy cells, but rather the death of the new workers that you need to power up your brain. It is therefore essential not only to boost your rate of neurogenesis but also to take action to keep your new brain cells from dying off—unless they're misbehaving.

The science in this area is evolving every day. We are still learning what kinds of things we can do to help grow new brain cells and discourage early cell death. But even now, we do know that there are certain lifestyle choices and environmental factors that affect your rate of neurogenesis and the survival of your newly formed neurons. Here are a few of them:

- **Environmental Toxins.** Exposure in your environment to neurotoxins, such as heavy metals, solvents, additives, or naturally occurring toxins from Mother Nature, will slash your rate of neurogenesis and kill off existing brain cells. Many neurotoxins can also change how your neurons use neurotransmitters, chemical messengers that are essential for maximum brain function. Mitotoxins are another type of toxin that harms mitochondrial performance, and these can kill your neurons because your neurons are extra sensitive to fluctuations in energy.

  We're swimming in a sea of neurotoxins every day. Now they're even added to our food. They are all around us—and inside of us. We'll talk in detail about how to both avoid neurotoxins and detox from them once they're in your body. For now, it's important to know that avoiding these toxins as much as possible and taking measures to enhance your body's detoxification systems are both key to boosting your brain performance.

- **Diet.** Perhaps the most important factor in determining your rate of neurogenesis is your diet. You can't make healthy, viable neurons without the right raw materials. Certain foods decrease your rate of neurogenesis while others enhance it. Two foods that slow the rate of neurogenesis are sugars and oxidized (damaged) fats. When oxidized fats get into your bloodstream, they cause inflammation. That inflammation slows your ability to make precious ATP, chews up the insides of your blood vessels, inhibits blood flow to the brain, and slows neurogenesis to a crawl.

  A high-sugar diet slows your rate of neurogenesis by increasing the amount of insulin in the bloodstream. Too much insulin degrades every organ in the body, including the brain. In one study,[18] mice that ate a high-sugar diet for just two months exhibited a significant decrease in cognitive function. Notably, the area of the brain that suffered the most was the hippocampus—where neurogenesis occurs. Sugar is the enemy of neurogenesis.

  There are other foods that can increase your rate of neurogenesis and keep new neurons healthy and active. Omega-3 fatty acids have a particularly profound impact on our rate of neurogenesis. A full one-third of the fat in our brains is docosahexaenoic acid (DHA), an omega-3 fatty acid, and studies have shown that increasing the amount of omega-3 fatty acids in your diet can promote adult neurogenesis.[19]

- **Bioflavonoids.** Plant compounds found in citrus fruits and many vegetables that are essential to keep these new neurons alive. So is a group of plant chemicals called polyphenols, which are found in coffee, chocolate, blueberries, grapes, and other blue, red, and orange foods. In fact, coffee fruit, the part of the coffee we usually throw away, has some of the best-studied polyphenols for growing neurons. Polyphenols are like Miracle-Gro for neurons.[20]

- **Stress and Depression.** Chronic stress has been shown to severely inhibit neurogenesis in the hippocampus.[21] Meanwhile, both stress and depression have been shown to cause neural atrophy and loss of neurons in the same part of the brain.[22]

Interestingly, antidepressant medications have the opposite effect, boosting the rate of neurogenesis in chronically depressed patients.[23] Some scientists now believe that the success of antidepressants is in part due to their impact on neurogenesis. The more new brain cells a depressed person is creating, the better he or she is likely to feel. The better a depressed brain can make energy, the better the depressed person is likely to feel. And as depression lessens, the rate of neurogenesis likewise improves. This is not a recommendation for more antidepressants—it's a recommendation for more neurogenesis.

For those of us who are not chronically depressed, it is helpful to know that anything we do to avoid chronic stress will also help us create more brain cells. On the other hand, temporary acute stress—stress that comes and then goes in a short period—can help tell your body that it's time to be more resilient and build new brain cells.

- **Exercise.** Exercise boosts your rate of neurogenesis by increasing the blood flow to your brain and putting your body through short-term healthy stress. It also triggers the release of nerve growth factors that protect new neurons against death.[24] Don't worry—you don't have to become a slave to the gym. The quick and easy exercise routines in this book are some of the best tools to grow new neurons and keep them alive. No CrossFit required—though you can do it if you like, and you'll perform better with more mitochondria and more neurons.

- **A Fun Environment.** Scientist Michael Kaplan[25] discovered that an enriched environment enhances neuron production in animals when he placed test animals in cages filled with interesting toys and monitored their levels of neurogenesis. I take this research so seriously that I designed the biohacking research facility where I wrote most of this book to be one of the most enriched environments I could dream of. It's full of interesting gadgets to keep those of us who spend a lot of time there engaged and entertained. If I'm going to go through the trouble of generating new neurons, I'll be damned if I'm going to let them die by spending days staring at beige cubicle walls!

- **Light and Water.** By now it's clear that anything that increases your mitochondrial function is also going to impact your neurons because they require so much energy. Intriguingly, studies show that mysterious cell structures called microtubules are a key part of shuttling mitochondria around in neurons. And recent breakthroughs in biology around the structure of water itself have illustrated how microtubules work.

  I had the good fortune to interview Dr. Gerald Pollack, bioengineering professor at the University of Washington. Dr. Pollack is editor of the scientific journal *Water*, and he discovered a phase of water that is not liquid, gas, or solid. This form of water, called exclusion zone water or EZ water, is critical to mitochondrial function and specifically to movement within microtubules. You get EZ water when you drink raw vegetable juices, fresh spring water, or glacial meltwater, and it forms spontaneously when regular water is exposed to infrared light or vibration. Better yet, EZ water forms in your cells when you expose your skin (and eyes, the gateways to the brain) to unfiltered sunlight for a few minutes every day without sunglasses or clothing or sunscreen. Low-level light therapy and infrared saunas can also have the same effect.

- **Sex.** In 2010, scientists conducted a study to observe the effects of sexual experience in the hippocampi of rats.[26] They did this by exposing adult male rats to a "sexually receptive" female either just once or once a day for fourteen consecutive days. In addition to looking at how their sexual experiences impacted their rate of neurogenesis, the researchers also studied the rats' stress levels when faced with the prospect of sex.

  The findings were incredibly interesting and telling for us humans. The rats that had sex just once experienced an increase in cortisol, the stress hormone, along with an increase in the number of new neurons in the hippocampus. The rats that had sex fourteen days in a row, on the other hand, did not experience the same increase in cortisol after the first day, but they did continue to see an improvement in their rate of neurogenesis.

How do these findings translate in our own bedrooms? Well, let's just say that even when you think you're too stressed for sex, it's a good idea to try to get in the mood anyway. This study offers compelling evidence that sexual experience can enhance our rate of neurogenesis while buffering us against the harmful effects of a short burst of cortisol. Your new brain cells (and your partner's) will thank you.

When you combine a high-fat, low-sugar diet that mimics fasting but will never leave you hungry, enhanced detox strategies, an exercise plan, and stress-relief protocols, you'll find that you can actually feel the Head Strong program supercharging your brain as it makes new neurons along with strong, healthy myelin. Better yet, these brain cells will keep working as they become seamlessly integrated into your smarter, faster, happier brain.

You're on your own when it comes to sex, though.

Head Points: Don't Forget These Three Things

- Your neurons require massive amounts of energy to function and they die when they don't get it.
- Myelin is made of fat and insulates the communication path between neurons.
- You can grow new neurons (neurogenesis) at any age, and it is possible to increase your rate of neurogenesis by five times!

Head Start: Do These Four Things Right Now

- Eat more good fat, particularly healthy saturated fat from grass-fed butter and meat.

- Eat a lot less sugar. Sugar reduces your rate of neurogenesis.
- Manage your stress. Stress reduces your rate of neurogenesis.
- Have more sex, but if you're a man, don't ejaculate every day because you'll waste hormones that are hard to make, along with precious mitochondria.

4

INFLAMMATION

The Muffin Top in Your Brain

Before I hacked my mitochondria, my brain didn't work as well as it should, I was often stupendously cranky for no reason, and I would go through random periods of having zero energy. Oh, yeah, and I was fat—I weighed about 50 percent more than I do now. But I also experienced an array of other physical annoyances that just seemed like a normal part of life because they had been lingering for years—sore joints, upper-back pain, blisters on my feet, countless canker sores, and a double-roll muffin top around my middle that seemed to shrink or expand from one day to the next. But I could still kick some ass, at least often enough to force my way through my career and relationships using willpower.

I didn't know at the time that inflammation plays a key role in mitochondrial performance. It wasn't until I began to focus on how I could increase my energy and mental focus that I discovered that all of these other symptoms were actually manifestations of the same problem: inflammation.

Inflammation itself is an important and useful physiological response. When a pathogen, toxin, or trauma stresses the body, a short-term burst of inflammation is part of its effort to protect and heal you. Just the other day, I had the chance to see inflammation work its magic on my seven-year-old son. While skateboarding, he fell down and hit his knee. Within seconds, his knee swelled to the size of a softball. That's because inside his body, blood, water, and white blood cells rushed to the site of the injury to begin repair work. Because my son is healthy, the swelling went down over the next few hours as the fluids drained from the tissue, the torn tissue began to heal, and a scab began to form over the damaged skin. All of that hard work required the energy produced by mitochondria!

This is an example of short-term or acute inflammation, which is healthy and necessary. Without acute inflammation, small cuts and scrapes would become life-threatening emergencies and you'd never be able to build new muscle from exercise (weight-bearing exercise induces mild, acute inflammation). Inflammation only becomes a problem when it is chronic, meaning that it lasts for a sustained period of time. With chronic inflammation, instead of specific areas of the body becoming temporarily inflamed so you can heal, your full body becomes inflamed and remains inflamed—indefinitely. Nobody enjoys feeling bloated and puffy all of the time, but the dangers posed by chronic inflammation are much more serious than simply not being able to fit into your favorite jeans. The odds are high that you have some degree of chronic inflammation slowing you down right now, but rest assured that the Head Strong program will help you address that. Less chronic inflammation equals better performance. It's that simple.

Inflammation manifests itself very differently in each area of the body, which is why it's not obvious that skin problems like acne and rashes may have the same underlying cause as your forgetfulness and fatigue. But chronic inflammation is also at the root of many life-threatening diseases. Together, cardiovascular disease, cancer, and diabetes account for almost 70 percent of all deaths in the United States. Alzheimer's and autism are growing at unprecedented rates. What is the common link between all of these diseases? Inflammation.[1, 2]

Your brain is actually the first part of the body to suffer when you are chronically inflamed. This is because your brain is sensitive to inflammation anywhere in the body. It doesn't matter if the source of inflammation is your heart, stomach, or left pinky toe—if any part of your body is inflamed, it will release chemicals called cytokines that negatively impact your brain. I didn't realize for a long time that the muffin top around my waist mirrored the muffin top in my brain.

Inflammation is at the core of most age-related neurodegenerative diseases. In the case of Alzheimer's disease, inflammation kills off neurons, causing memory loss and other cognitive problems,[3] although there are clearly other mitochondria-related parts of the disease, too. The prefrontal cortex (your human brain) is even more sensitive to inflammation than the rest of your brain, which is why even "normal" aging is associated with decreased cognitive function and forgetfulness. These symptoms, which are commonly dismissed as by-products of aging, are actually symptoms of inflammation,[4] which itself lowers mitochondrial function.

Scientists have established a connection between non-pathological (or "normal") inflammation in the aging brain and cognitive decline in many species, including pigeons, rodents, and humans.[5] Inflammatory genes are found in higher quantities in aging brains than in youthful ones.[6] Aging brains also have an exaggerated response to stress and infection,[7] meaning they become inflamed more easily.

Well, you already know that "normal" brain aging is for suckers. But you don't have to be old to be suffering from brain performance issues caused by inflammation. Studies show that at any age, inflammation hurts brain performance, specifically your ability to learn, remember, and pay attention.[8] In multiple studies, mice injected with inflammatory substances showed evidence of cognitive impairment in the areas of the brain that govern special relations, learning, and memory.[9] Interestingly, it didn't seem to matter whether the inflammatory substances were injected directly into the brain or into another part of the body. Any inflammation ultimately led to cognitive decline.

This all means that the amount of inflammation in your body is impacting your brain's ability to think, learn, and remember *right now*. This is not just something that will eventually catch up to you. You're unfocused *right now* because you're inflamed. You can't remember things *right now* because you're inflamed. You're not as sharp as you want to be *right now* because you're inflamed. And the worst part is, you probably don't even know it. Unchecked inflammation takes away your mental edge long before it causes you any physical pain or discomfort. This means that if you're already experiencing some physical signs of inflammation, your brain has most likely been weakening for a long time.

When you dial down the swelling in your body or in your brain, the fog suddenly clears. You can think more clearly, focus when you want to, and remember things with ease. A huge portion of the Head Strong program is dedicated to reducing inflammation throughout the body and therefore in the brain. This is one of the improvements that will give you the fastest and most noticeable results, in large part because inflammation makes your mitochondria work less effectively.

When I was getting my MBA at Wharton, I remember studying extra hard for an exam in a quantitative finance class where I was really struggling. By the morning of the exam, I felt so prepared! But before the test I ate a salad with a bunch of overripe avocados that had started to go bad. Within a few minutes, I felt my energy reduce. I remember wondering how it was possible that I was starting to get blisters on my feet after walking just a quarter mile to class. Unbeknownst to me, the toxins in my food had triggered inflammatory chemicals to start raging around my body and my brain. When I sat down to take the test, it didn't matter how hard I concentrated—it was like I hadn't studied at all. Ultimately, I failed the exam and barely passed the class. At the time, I had no idea that my brain performance had been limited by the same thing that was making my feet hurt—inflammation.

WHAT CAUSES INFLAMMATION?

Anything that irritates the body, including any form of physical or psychological stress, can cause inflammation. Your body's inflammatory response is similar to your fight-or-flight response. It is there to keep you alive, in this case so you can heal from an infection or an injury. This response only becomes problematic when it is turned on more often than is necessary or healthy—which unfortunately is the case for most of us. We get less sleep than we should, we are exposed to a daily soup of environmental toxins and light pollution, and we eat processed foods with ingredients that are completely foreign and irritating to our digestive tracts. All of these factors stress our bodies and cause inflammation, which therefore reduces our mitochondrial function.

Our contemporary Western diet, with its huge amounts of inflammatory foods, creates an inflammatory internal environment, either because of chemicals that cause inflammation by harming mitochondria or because our immune system constantly mounts a response against a perceived enemy. In the first instance, when you ingest something that stresses or irritates the gut, such as a toxin or pathogen, it triggers the immune system. Your body responds as if it's been injured and becomes inflamed in an effort to protect itself. Then it gets even worse—your irritated intestinal lining can develop microscopic tears that allow undigested food particles and bacteria to enter your bloodstream. A system-wide inflammatory response is triggered as your body attacks these foreign particles.

In the second instance, when your immune system responds to an enemy (real or imagined), it does this by releasing a stream of small inflammatory proteins called cytokines into your bloodstream every time it detects a foreign intruder. These cytokines travel through your body, causing oxidative stress to your cells, particularly the mitochondria. In a short amount of time, these cytokines enter your brain, causing an inflamed brain that simply can't make energy the way it's supposed to.

The degree to which your brain is inflamed plays a much bigger role in how you feel from day to day than you probably realize. An inflamed brain can make you unreasonably angry, can be a source of severe food cravings that distract you from whatever you're trying to accomplish, and can rob you of your memory, like it did to mine during that exam. Depending on your genetics, your body may respond to inflammation by mounting an autoimmune reaction, which causes even more damage as your immune system attacks important systems in your own body.[10]

It doesn't stop there. Inflammation can break down the energy production systems in your cells. Remember the electron transport system? It's how your mitochondria create energy by moving electrons around. In chapter 2 we discussed that the more efficient your body is at moving electrons in your mitochondria, the more energy and less inflammation you have as a result. But what happens when inflammation causes the cells themselves to swell? The distance those electrons have to move increases, and your mitochondria have to work harder to make the same amount of energy.

That means if you're inflamed, your neurons won't work as well, since they're so dependent on mitochondria. And when your mitochondria aren't producing energy efficiently, they make a lot more free radicals, which themselves cause inflammation and aging. Yikes!

INFLAMMATION AND THE GUT

We now know that roughly 50 percent of the body's immune system is clustered around your digestive tract. This may sound strange, but our bodies were designed this way for a very good reason. Dr. Jeffrey Bland, one of the founders of the field of functional medicine, told me during his interview on *Bulletproof Radio* that throughout the course of my life I would eat about twenty *tons* of foreign molecules. You're no different, and your body needs a way to translate those foreign molecules into messages it can understand so that it can determine which ones are friends and which are enemies.

The immune cells in your gut signal an alarm when they come across something they believe is an enemy. This alarm produces inflammation throughout the body to deal with the threat—which you may experience as bloating or brain fog. Some foods, such as anything fried or charred, contain chemicals that directly cause inflammation by harming mitochondria, while other foods only affect some of us. (Because of our particular mitochondrial DNA, nuclear DNA, or environmental exposures, some people are more sensitive to certain foods than others.) These mild, systemic inflammatory responses are not the same thing as acute allergies that cause immediate rashes and anaphylactic shock. The type of inflammatory response I'm describing here is the result of a food "sensitivity" that affects mitochondria. Cumulatively, exposure to such foods can ignite a high level of inflammation in the body.

The bacteria, or microbes, in your gut also play a role in determining how your body responds to certain foods. If the immune system in your gut is translating the messages in your food, it's important to have gut bacteria that understand the right languages. Unfortunately, many of us have wiped out many strains of our gut bacteria with antibiotics, pesticide-tainted foods, bad fats, and processed foods. The more we traumatize our gut bacteria, the more we threaten the delicate balance of our immune systems and the more likely we become to suffer from chronic inflammation.

When I interviewed Dr. David Perlmutter, a renowned neurologist and the author of *Brain Maker* and *Grain Brain* (both books I highly recommend!), he shared a fascinating study that compared the variety of bacteria present in the gut and the presence of parasites in humans to their rates of Alzheimer's disease. The people who had fewer types of gut bacteria and more parasites had higher risks of developing Alzheimer's.

How can parasites or the wrong gut bacteria lead to Alzheimer's disease? Well, until a couple of years ago, scientists believed that the brain was sealed off and protected by the blood-brain barrier, which separates circulating blood and particles from the brain. But we now know that gut bacteria help control what can get through your blood-brain barrier. They do this by creating a short-chain fatty acid called butyrate that helps maintain the integrity of the blood-brain barrier.[11] When your gut bacteria don't make enough butyrate, the blood-brain barrier becomes more permeable, allowing particles into the brain that should be kept out. This, of course, leads to inflammation, as your body attacks the intruders. Don't worry—there are tricks to get your gut bacteria to make more butyrate, and it's easy (and delicious) to ingest it from its richest dietary source: grass-fed butter.

There is another intriguing angle to the story of how gut bacteria impact mitochondria and inflammation. We have learned that mitochondria make biophotons—tiny pulses of light that last for one-quadrillionth of a second. This is part of how they communicate with each other. As it turns out, gut bacteria also make biophotons. Is it possible that your gut bacteria are communicating with the tiny bacteria-derived mitochondria that rule your cells? I think it is, especially because we know mitochondria are sensitive to external sources of light. Further studies will help substantiate or disprove this idea.

Studies conducted on mice provide great insight into how the types of bacteria in your gut can influence your inflammation levels. When bacteria from the guts of mice with chronic inflammation and obesity were placed in the guts of skinny mice, the once-slender mice overate by 10 percent and became insulin resistant. On the other side of the coin, when bacteria from lean mice were inserted into the guts of obese mice, they became leaner.[12]

Like mice, obese people and lean people have very different populations of gut bacteria. Whether it's the bad gut bacteria that cause obesity or obesity that causes bad gut bacteria is still unknown,[13] but there is evidence that the wrong kind of bacteria in your gut causes insulin resistance and inflammation.[14] And we know that inflammation slows mitochondria.

Chronically inflamed people (and animals) often have an excess of bacteria from the firmicutes family, which includes the lactobacillus bacteria found in yogurt and most probiotic supplements. You need these bacteria—in fact, they are the most common ones in your gut—but if you have too many of them compared to another class of bacteria called bacteroidetes, you'll likely have more inflammation. Naturally thin people have fewer firmicutes and more bacteroidetes.

You can't buy bacteroidetes species probiotics as a supplement, but you can easily generate them by eating foods that contain their natural food source, polyphenols. You read in chapter 3 that polyphenols are antioxidants that help keep newly formed neurons alive. It turns out that polyphenols also do wonderful things for mitochondrial performance, including reducing free radicals and even causing you to grow new mitochondria.[15, 16] Since they also feed the good bacteria in your gut, adding more of these plant compounds to your diet or supplementing with them is a no-brainer (ha!).

When I learned about the connection between the gut and the brain, I was determined to hack my gut microbiome. I already knew that I'd traumatized my gut as a kid, when I took antibiotics about once a month for the chronic strep throat infections I had for years. So I started experimenting with every fancy probiotic I could find. I even found a company in Thailand that bred eggs from a parasite called a porcine whipworm. I ordered the eggs and swallowed them so they would hatch in my gut, where they would live for only six weeks. This is called helminth therapy, and for some people, this treatment heals the gut and leads to a dramatic reduction in inflammation throughout the body. It didn't work for me, but every two weeks while writing *Head Strong*, I swallowed up to sixty rat tapeworm larvae called

HDCs (*Hymenolepis diminuta cysticercoids*) as part of an ongoing allergy elimination protocol to reduce inflammation. Another guest on *Bulletproof Radio*, Dr. Sidney Baker, has pioneered this method of reducing inflammation. This may sound pretty extreme—and it is. Don't worry, though—this is not part of the Head Strong program. But I think it's exciting to be at the forefront of discovering more about the brain-gut connection.

Despite trying all of these radical techniques, I've noticed the greatest improvement in my performance when I simply add more butyrate-rich grass-fed butter and polyphenol-rich foods to my diet. Eating more of these superfoods has helped me maintain my gut lining while reducing the inflammation in my brain.

We are learning more about the connection between the brain and the immune system every day. Just recently scientists at the University of Virginia School of Medicine published a groundbreaking discovery showing that the brain directly connects to the immune system via lymphatic vessels that we didn't previously know existed.[17] The researchers say the vessels were easy to miss because of their location, closely shadowing a major blood vessel down in the sinuses, an area that is difficult to see in standard medical imaging scans.

Until this study was published, medical schools had long taught that cytokines—the molecules of inflammation that harm the brain—could only impact the brain after they penetrated the blood-brain barrier, which was supposed to be impossible since we thought these molecules couldn't pass the barrier. We now know there's a direct line between the immune system and the brain. We also have a plausible explanation for how food can dramatically impact cognitive performance, as I learned firsthand that day when bad avocados sabotaged my test-taking skills.

Though this is still an emerging field, the evidence exists now to show a clear link between the foods you eat and your mental performance. When you ingest something that triggers an inflammatory response in your body, either because it's toxic to your mitochondria or because you're sensitive to it, your mitochondria pay the price and your brain will suffer. When the inflammatory response calms down, your mitochondria can recover, and your brain will be able to function at full speed, perhaps for the first time in your life.

INFLAMMATION AND HORMONES

Your body's hormone function has a wide-ranging impact on inflammation. We know that hormone dysfunction causes inflammation, and inflammation causes hormone dysfunction. It's a two-way street, and it can become a vicious cycle. But the good news is that certain hormones actually protect us against inflammation—sometimes the same hormones that can become inflammatory under different circumstances. For example, testosterone is an anti-inflammatory[18] hormone, while estrogen is sometimes anti-inflammatory but is often pro-inflammatory.[19] As you read in chapter 3, progesterone is a hormone present in male and female brains and is required for normal neuron development. Doctors are even using it to help treat traumatic brain injury because it helps prevent neuron loss and regulates inflammation. I took advantage of this effect when I got a mild concussion while writing this book. With a physician's advice, I took progesterone for a week after the concussion and saw very noticeable improvements in inflammation and cognitive function.

One of the most important and under-recognized hormones when it comes to your body's inflammatory response is called vasoactive intestinal polypeptide (VIP). Your body makes VIP in your gut, pancreas, and two important parts of your brain, the pituitary gland and the hypothalamus. VIP protects against inflammation, controls and sends nerve signals, triggers the release of other hormones, and improves brain function, sleep, and glucose control. VIP also has a hand in regulating learning and memory, immunity, and responses to stress and brain injury.[20]

In short, VIP is critical to proper brain function. When your body is under physical or psychological stress, it makes less VIP than usual and your inflammation rises accordingly. We've seen in studies that when mice are exposed to certain toxins—in this case a toxic mold commonly found in food—their VIP levels drop.[21] We also know that humans experience a similar drop in VIP when exposed to environmental toxic mold, which is a problem I suffered from as a young man and which is a common problem in the United States, according to a dozen experts I interviewed for the documentary *Moldy*. Did mold exposure make me stupid because it dropped my VIP level, because it increased my body's inflammatory response, or because it damaged my mitochondria? I'd bet on all three!

Animal studies show that when animals don't have enough VIP, their blood sugar and insulin levels rise, and the animals crave sweets.[22] The same thing will happen to you when your lifestyle—or a toxin—lowers your VIP. You'll likely reach for a sugary snack. You may think that eating a little junk food is no big deal, but it actually triggers a widespread inflammatory response in the body and lowers the production of the very VIP that you need to protect you from that inflammation. This is how you can get caught in a self-amplifying cycle of brain fog, fatigue, and, most likely, regret. The Head Strong program is designed to control this little-known hormone so you don't get caught in this harmful cycle.

Another substance that affects your brain (and muscles) has the very unsexy name mammalian target of rapamycin, which is why it usually goes by mTOR. It's not technically a hormone, but it plays an important role in controlling inflammation by regulating cell growth, cell survival, and cell death (autophagy). By now, you know that mitochondria really control those functions in your body, so mTOR must work with them. Having a healthy balance of this substance is key. Too much mTOR is a bad thing, as it contributes to inflammation and increases your likelihood of developing cancer, obesity, and neurodegenerative diseases. But if you have too little mTOR, you'll also miss out, because it increases energy production in your mitochondria and encourages new mitochondrial growth.[23] Better yet, mTOR helps improve your memory,[24] and occasional spikes of it help you grow muscle.

Studies have shown that following a calorie-restricted diet or one that mimics fasting suppresses mTOR, which leads to an increase in the type of cells that help fight inflammation.[25] The Head Strong program shows you how to use simple strategies involving coffee, exercise, and diet so that you never have too much mTOR but you occasionally "pulse" your levels of mTOR to keep your mitochondria running like race cars. These same techniques help you easily maintain the "almost muscular" look that helps you live the longest.

SNEAKY FATS

Another class of molecules we can control to help lower inflammation is called eicosanoids. These molecules act as messengers in the central nervous system, triggering an immune response after you ingest something your body perceives as toxic. If you've heard of eicosanoids, it's probably due to the work of Dr. Barry Sears, creator of the famous Zone Diet. When I interviewed him, he highlighted how eicosanoids can make your brain function for better or worse.

Your body makes eicosanoids from either omega-3 or omega-6 essential fatty acids. The eicosanoids made from omega-6s are pro-inflammatory, while the eicosanoids made from omega-3s are anti-inflammatory. Our bodies need a balance of both types of fats in order to produce both types of eicosanoids. Remember, our bodies need to be able to become inflamed in times of trauma or after exercise, so it's important to have some omega-6s and pro-inflammatory eicosanoids.

The problem is that balance can be difficult to achieve if you're eating a standard American diet, which contains far more omega-6s than omega-3s. Vegetable oils, which are a main source of omega-6s, have become the cheapest source of fat calories in the world. You will find vegetable oils in almost every packaged food at your local grocery store or restaurant meal. As a result, our consumption of omega-6 fats has dramatically increased over the last fifty years. Since the omega-6s are the building blocks of pro-inflammatory eicosanoids, our levels of inflammation have risen along the same trajectory.

Is it any wonder, then, why we've seen such a steady increase in the incidence of cognitive and neurological dysfunction and inflammation-based diseases? Today, most older Americans are concerned about their risk. But if you are diagnosed with Alzheimer's at age eighty-five, the damage really began to set in at least thirty years earlier. That's when you had the chance to dial down the inflammation throughout your body. Hopefully for you, that time is now. It's far easier to prevent decline before it happens than to try to fix your body after it's already worn out—especially when it comes to diseases for which there is not yet a cure.

While we've steadily increased the amount of inflammatory omega-6s in our diets, we've also continuously upped our intake of sugar. In the year 2000, the average American consumed fifty-two teaspoons of added sugar a day.[26] This is like adding a lighted match to a vat of gasoline: the result is a giant explosion of inflammation.

Eating sugar causes insulin levels to spike, which triggers the release of inflammatory cytokines. Fructose (which makes up 50 percent of table sugar) also easily links to proteins in the body such as collagen, the main constructive tissue in our skin and arteries. When linking to collagen, fructose creates toxic advanced glycation end products (AGEs). These are aptly named, as these end products play a role in the aging process and create oxidative stress in the body,[27] triggering even more inflammation.

The oxidative stress caused by AGEs damages mitochondria, and there is a direct relationship between inflammation and mitochondrial dysfunction.[28] Once again, this is a two-way street. Poor mitochondrial function leads your body to create molecules that cause inflammation, and these molecules then damage your mitochondrial function even more. And that leads to sugar cravings!

When you take into consideration all of the ways that inflammation affects the body, it is clear that many chronic diseases start out looking very similar on a cellular level. Diabetes, heart disease, Alzheimer's, and other degenerative diseases are all the result of mitochondria problems.[29] It's also clear that nearly every cause of lowered energy production in your body *now* lays the groundwork for progressive declines in your performance and the development of chronic diseases *later*. Most people have some under-performing mitochondria, too much oxidative stress, and chronic inflammation. The mitochondrial toxins you eat or are exposed to every day rev up your body's inflammatory response and inhibit neurogenesis and mitochondrial function. That doesn't guarantee that you'll develop an autoimmune condition, a mental health problem, or a neurodegenerative disease like Alzheimer's or MS. But the origins of these diseases are all surprisingly similar to the symptoms we associate with simply "having a bad day."

Head Strong is all about putting you in control of how good every day will be from now on. And it turns out that you can kill two birds with one stone because the exact same things that prevent and reverse chronic diseases have also been linked to an increased level of brain function and performance. A level of performance that I had never known was possible. A level of performance that you can begin to experience immediately.

And it all starts with your mitochondria. But first, there's one more thing on the inflammation front . . .

INFLAMMATION, WATER, AND LIGHT

Water is intimately involved in everything our cells do. Our cells are two-thirds water by volume, but those water molecules are so small that ninety-nine out of every one hundred molecules in your cells are water molecules. The amount and type of water in your cells is incredibly important to your every bodily function.

In chapter 3 you read about Gerald Pollack and his discovery of a fourth phase of water called EZ water. It is this type of water that is inside your cells. When you don't have enough EZ water in your cells, they become dehydrated and stop functioning well. Your body's lymphatic flow (which includes toxins and waste products) also becomes inhibited, leading to chronic inflammation. Your mitochondria need EZ water. In fact, they can't function without at least some of it.

The EZ water inside your cells is negatively charged. This is important because, as we learned in chapter 2, neurons must be negatively charged before they send a signal to communicate with another neuron. If your cells don't contain enough EZ water, they aren't as negatively charged as they should be, and your neurons can't communicate efficiently. Poor communication between the cells in your brain can lead to all sorts of cognitive problems as well as depression and other mood disorders, even the minor ones that make you yell at your kids when you really didn't mean to.

The more negative charge you have in your body, the better all of your cells are able to function. Oxidation causes cells to lose their negative charge. Antioxidants fight excessive oxidation by trying to preserve that negative charge. Your body actually has many ways of holding on to a negative charge, including urinating, sweating, exhaling carbon dioxide, and even defecating. All of those waste products are positively charged.

One thing I have done for nearly a decade to increase my body's negative charge is called earthing, or soaking up negative charge from the ground. This can help your body build more EZ water. Flying is a mode of travel that tends to reduce your negative charge, which lowers your amount of EZ water and causes inflammation. This is one reason you experience jet lag. Years ago, when I spent a year commuting from San Francisco to Cambridge, England, every month, I noticed that I performed *much* better if I did yoga in a park barefoot after I landed. At the time, I had no idea why this worked, but now I know that I was soaking up negative charge from the earth and helping my body build EZ water. Of course, the mental and physical benefits of the yoga and breath work shouldn't be discounted, either. But I noticed a marked difference when I practiced yoga barefoot on the ground. Four years later, a paper was published in the *Journal of Environmental and Public Health,* explaining some of the grounding effects I'd experienced.[30]

Light therapy is another way to help your body make EZ water. When regular water is exposed to infrared (and maybe UV [ultraviolet]) light, it can be transformed into EZ water. If you expose yourself to infrared light via infrared sauna or simply by going outside on a sunny day without sunglasses or sunscreen, your body will soak up that light energy and build EZ water. Light enters your body through your eyes and makes its way directly into your brain, where you'll first feel its impact. Light matters greatly to your brain, in part because of its ability to help make EZ water.

Dr. Pollack described an experiment in his lab in which he flowed water through a narrow tube. When he exposed that water to UV light, it flowed through the tube five times faster. If your blood and lymphatic fluid can flow through your narrow capillaries more quickly, you will experience less chronic inflammation. The tiny microtubules in your mitochondria also benefit from this "turbocharged" effect when you are exposed to sunlight.

After I first came across this research from Dr. Pollack and Nick Lane, a British biochemist who has studied light, I slowly increased the UV light exposure to my eyes, and I found it had a profound, noticeable effect on my cognitive function on the days I did it. This was likely due to the increase of EZ water in my neurons and a decrease in overall inflammation.

For some time, NASA—the National Aeronautics and Space Administration—has been studying the effects of light to speed healing, soothe sore muscles, ease chronic pain, relieve stiffness, and increase circulation. Light exposure can do this by improving mitochondrial efficiency and protecting against inflammation.[31] Light therapy has also been shown to help mitochondria make ATP faster.[32]

Nearly twenty years ago, I was in a car accident and got whiplash for the second time in my life. The first time I had it, it took me nearly a year to recover. And after I was rear-ended for a second time, I was in a lot of pain. A naturopath friend met me in a parking lot in San Jose, California, and handed me a small handheld medical laser. He told me to put it on my upper back where I was injured. Within three minutes of this laser pulsing red and infrared light into my back, my pain was gone. The relief was faster than anything I'd ever experienced. I bought the device, and I've been pursuing light therapies ever since.

Today, I use the REDcharger, which has more than forty thousand red and infrared LED lights that illuminate my entire body to help me recharge my mitochondria, make more EZ water in my cells, and grow healthier collagen. It's an amazing biohack, and you can use red light at home very affordably. Later in the book you'll read my specific recommendations for light exposure. This is an exciting area that is not often discussed but that can have a dramatic and immediate impact on your mood, your inflammation level, and your cognitive function.

By offering exposure to the right kinds of light, feeding the good bacteria in your gut, balancing your hormones, and of course improving your mitochondrial function and efficiency, the Head Strong program will help you dial down the inflammation that is quietly raging throughout your entire body—and especially in your brain. Your brain's lumpy muffin top will be replaced by a streamlined, fast-acting machine that you can count on to help you be your best. This is the joy of an uninflamed brain.

Head Points: Don't Forget These Three Things

- Your brain is the first part of the body to suffer when you are chronically inflamed.
- When your mitochondria are inflamed, they are less efficient at making energy because electrons have to travel farther to get to the same place.
- Nearly every cause of lowered energy production now lays the groundwork for developing chronic diseases later.

Head Start: Do These Three Things Right Now

- Pay attention to changes in your muffin top; the same foods that cause your body to become inflamed are also making your brain foggy and inflamed.
- Spend some time outdoors barefoot to soak up the earth's negative charge and get some UV and infrared light. If it's winter where you live, use an earthing mat and a sauna.
- Have your inflammation levels checked by a functional medicine doctor. Substances such as CRP (C-reactive protein), homocysteine, and Lp-PLA2 (lipoprotein-associated phospholipase) are good markers to review for inflammation.

PART II

YOU ARE IN CONTROL OF YOUR HEAD

Now that you understand some of the signs that your head is *weak*—like forgetting things you want to remember and suffering from annoying mood swings, low energy, food cravings, and a lack of focus—it's time to get in control and become Head Strong. You have the capacity to improve your mental abilities and sharpen your focus—and the tools are closer than you might imagine.

I wish I'd known that I could influence my brain performance all those years ago. I had no idea how many of my symptoms were within my control; I had to wait until my brain hit rock bottom before I learned that it was possible to improve my mental ability. In fact, I was so scared that my brain would never work again that I actually bought disability insurance when I was twenty-six.

Luckily for all of us, the brain is highly adaptable. In fact, this is one reason we can hack the brain—because it responds to even subtle changes in our bodies or the world around us. The other factor that makes the brain hackable is, ironically, its complexity. Hackers like complex systems because they have large "attack surfaces"—lots of opportunities to intervene and take over. The simpler the technology, the more difficult it is to hack the machine. Simple machines may be easy to understand, but they offer fewer entry points and therefore fewer opportunities to intervene and influence their performance.

Complex systems may be easier to hack, but they're also tougher to understand. Our biology is *really* complex—it's a mix of chemical, electrical, physical, light, and magnetic signaling. It's affected by temperature and light, season, time of day, and many other small variables.

The brain—and especially its energy-making process—is so complex that there are multiple ways to speed it up. The good news is that if you skimmed (or even skipped!) to this point of the book, that's okay. You don't need a detailed road map of a system to hack it; you just need to believe it's possible and have the ability to change some of the systems' inputs. For us (and our brains), that's often just changing the environment around us.

And that's what Part II of this book is all about. You'll learn exactly which buttons to push and levers to pull to dramatically speed up your cellular energy production, start growing healthy new mitochondria and neurons, and reduce your inflammation. You'll likely feel these results from day one. Even better, these changes will compound over time and change your trajectory so that massive change becomes possible by making just a few small shifts. When you move to a bank that offers you a 2 percent higher interest rate, the result is not massive growth of your savings overnight but an exponential increase as that interest compounds over years. Similarly, these brain triggers and levers will each lend you an extra edge that, when combined, will dramatically change the way you think, feel, and perform every single day and become more powerful with time.

These are the energy secrets I used to create the power and focus to grow Bulletproof so that I could reach millions of people while working at a full-time job and being a new dad. In fact, I'm using nearly all of them simultaneously while writing this book for you! They work for me, and they will work for you. Now let's get to hacking and power up your brain.

5

BRAIN FUEL

When it comes to taking control of your brain, your diet offers the easiest and most powerful way to gain the upper hand. The food you put into your body has a greater impact on your brain's performance than any other single factor within your control. Nutrition is literally the fuel that either helps your brain run efficiently or causes it to break down. Your body takes electrons from the foods you eat and combines them with water and oxygen to make energy. You really are a battery! And you need the right materials to stay powered up.

POLYPHENOLS: BRAIN-BOOSTING ANTIOXIDANTS

As we've discussed, your mitochondria need oxygen to make ATP, but this energy-making process generates free radicals, and some of them leak out to cause havoc in your cells. If you don't want inflammatory free radicals harming your performance and making you feel old, you need to do what your grandmother told you—eat your vegetables. All vegetables contain antioxidants, but some are far more potent than others.

Polyphenols are a type of antioxidants found in plants—in particular they are found in the dark red, purple, and blue parts of plants. Not only do polyphenols help to protect us against cellular damage from oxidation but they also offer a bunch of protective properties for our mitochondria. Here are a few of the ways polyphenols can benefit us.

- **They Protect Your Gut.** Polyphenols can actually change the composition of your gut bacteria, increasing the amount of healthy bacteria and inhibiting the growth of harmful bacteria in your intestinal tract. In one study, people who drank a blueberry extract high in polyphenols for six weeks saw a dramatic improvement in their gut bacteria.[1] This type of positive change to your gut bacteria can help lower inflammation and reduce brain fog.

  Studies have also shown that polyphenols can protect your gut from dangerous pathogens such as staphylococcus and even salmonella.[2] Interestingly, polyphenols and your gut bacteria have a symbiotic relationship. Polyphenols change the composition of your gut bacteria, while your gut bacteria are responsible for metabolizing polyphenols so your body can use them.

  A study of rodents showed that those that were fed polyphenol-rich coffee (with butter, the same idea behind Bulletproof Coffee!) had a higher ratio of polyphenol-eating healthy gut bacteria than those that didn't, and this was associated with being leaner.[3]

- **They Increase Your Rate of Neurogenesis.** Studies have shown that polyphenols increase your levels of brain-derived neurotrophic factor (BDNF) and brain nerve growth factor (NGF).[4] These are both proteins that encourage neurogenesis and protect your new neurons from dying off. An increase in BDNF and NGF has also been shown to improve learning, memory, and thinking.[5] That's why raising the levels of these proteins is such a key part of the Head Strong program.

- **They Tell Your Cells Whether to Live or Die.** Polyphenols play an important role in facilitating the cellular signals that initiate the process of apoptosis (cell death) and prevent old or damaged cells from mutating.[6] You want cells that have already mutated to submit to apoptosis, but it's even better if those cells don't mutate in the first place. Polyphenols can help keep your cells—and you—strong, healthy, and alive.

- **They Help Fight Inflammation.** In addition to protecting your brain from inflammation by feeding the good bacteria in your gut, polyphenols reduce the amount of inflammatory cytokines in the bloodstream. This reduction in inflammation has been shown to improve blood flow to the brain,[7] which enhances memory and prevents age-related declines in cognitive function. And you've already learned that a decrease in inflammation is going to enhance mitochondrial function.

Now, you may be thinking—if polyphenols are so great and can be found so easily in food, aren't we all benefiting from their amazing properties already? Well, the thing about polyphenols is that, with some exceptions, your body can't easily absorb them. The majority of polyphenols need a little help for your body to actually use them. And the best way to do that is to consume them with my all-time favorite macronutrient: fat.[8] That's right, putting butter on your broccoli actually makes it *better* for you because your body can absorb more of the good stuff in the broccoli when those polyphenols hitch a ride with fat molecules. I have a pet theory that broccoli evolved all those little florets just to better absorb butter, even though there's no actual evidence of that.

On the Head Strong program, you are going to eat lots of fresh dark veggies and low-sugar fruits to get the recommended amount of polyphenols per day, which is at least 2 grams. The Head Strong recipes will show you how to combine the best food (and beverage) sources of polyphenols with enough fat to make them easily absorbed by your body. Within two weeks of consuming more polyphenols in a more potent way, you will experience less inflammation, a healthier gut, and a higher rate of neurogenesis.

So, in addition to dark-colored vegetables and low-sugar fruits, what else should you be eating to get as many protective polyphenols as possible? The following foods offer the highest and most beneficial concentration of polyphenols. Consuming more of them will help you become Head Strong by fighting inflammation, keeping your gut in balance, and helping your brain to grow healthy new brain cells.

- **Coffee.** If you're already familiar with Bulletproof Coffee, this will come as no surprise to you. If you're new to my world, welcome. Here, coffee is king. It is rich in polyphenols and contains over a thousand different compounds that improve the function of your cells.[9] Coffee is the number one source of polyphenols in the Western diet,[10] which is why it's at the top of the list. (And if something gross like sardine juice were higher in polyphenols that would be at the top of the list instead!)

  The polyphenols in coffee regulate the "switches" that turn certain genes on and off, including the one that signals cells to replicate or to die.[11] Coffee also contains a type of polyphenol called chlorogenic acid, which reduces chronic inflammation, particularly in cells with a high fat content, such as brain cells. This is one way (of dozens) that coffee improves cognition.[12]

  But perhaps the best way coffee can keep you Head Strong is by helping you live longer. A huge study found a strong association between coffee consumption and longevity. The more coffee the study participants drank, the less likely they were to die.[13] In fact, drinking coffee was associated with a lower risk of dying from a number of common diseases, including heart disease, lung disease, diabetes, and infections. (Did you notice that all of these are mitochondrial disorders?) And in women, the effect was about 30 percent stronger than in men.[14] I think it's possible that this stronger effect in women is related to the fact that some cells in women's ovaries have ten times more mitochondria than brain cells.

  Given that low mitochondrial function leads to all of these diseases, if the polyphenols in coffee improve mitochondrial function, it would follow that they also help to prevent these common diseases. And the study participants consumed both caffeinated and decaf coffee, so we know that caffeine can't be responsible for these effects.[15]

  You'll learn more about Bulletproof Coffee later on. For now, the important thing to know is that we pair toxin-free coffee (a lot of coffee out there contains mitochondrial toxins from mold) with fat (in the form of grass-fed butter and Brain Octane Oil) so you can better absorb the polyphenols and avoid the milk protein that blocks your body from using polyphenols.

- **Dark Chocolate.** Dark chocolate (at least 85% dark) is full of polyphenols and contains a mild amount of caffeine, which also enhances performance.[16] You have to be careful when selecting your chocolate products, though, because, just like coffee, a lot of chocolate contains mold toxins that actually inhibit mitochondrial function. All chocolate is produced by fermentation, and 64 percent of the microbes that ferment chocolate can also create harmful mold toxins.[17] European chocolate tends to be the lowest in mold toxins, as it is regulated under stricter standards, so I choose European chocolate if I don't have my own lab-tested ultra-pure chocolate on hand.

- **Blueberries.** Studies on the polyphenols in blueberries show that they increase life span, slow age-related cognitive declines,[18] and significantly improve cardiovascular function.[19] They directly raise BDNF, too! Unfortunately, blueberries are heavily sprayed with pesticides and are often visibly moldy, especially bulk frozen berries. I recommend choosing high-quality organic frozen or farm-fresh blueberries or using a purified extract of blueberry polyphenols; if you ate enough blueberries to get all of the polyphenols you want, you'd be getting a ton of sugar, too.

- **Pomegranates.** The type of polyphenols found in pomegranates is water soluble and more easily absorbed by the body than most other polyphenols. These polyphenols are also known to break down into smaller compounds that cross the mitochondrial membrane and fight oxidative stress directly in the mitochondria.[20] When your gut bacteria digest pomegranate, they make urolithin A, one of the few substances that cause your body to replace worn-out mitochondria with fully functioning new ones. Fresh pomegranates (and their juice) have a special detox enzyme called PON1 (paraoxonase 1), but pomegranate juice is too high in sugar to consume regularly. I eat fresh pomegranates whole when they're in season and use an extract the rest of the time.

- **Grape Seeds.** Grape seeds contain a very powerful type of polyphenol called proanthocyanidin. You don't need to remember the name, just that it's different from the polyphenol found in red wine, resveratrol. Grape seeds—and particularly grape seed extract—have potent anti-inflammatory properties[21] and protect your brain from oxidative stress.[22] But the real reason I'm excited about grape seed extract is that animal studies have shown that it corrects mitochondrial dysfunction caused by obesity and it protects against weight gain.[23] Researchers in one study described grape seed extract as capable of correcting an energy imbalance and improving the fat-burning capacity of brown fat. Brown fat is the type of fat that helps you make energy and burn fat. In fact, it is brown instead of white because it is so rich in mitochondria.

  In another study, scientists pretreated animals with grape seed extract before inducing massive brain stress. The pretreatment reduced oxidative stress, mitochondrial free radicals, and neuronal and mitochondrial damage.[24] To stay Head Strong, you want to limit mitochondrial damage, too! That's why I take grape seed extract every day. In my teens, grape seed extract was the first supplement I took that had a noticeable effect. I'd been struggling with daily nosebleeds—a symptom of inflammation from living in a water-damaged building with toxic mold—and grape seed extract reliably reduced my number of nosebleeds. I went from getting several nosebleeds a day to rarely ever getting one. This was a game changer for my dating life at the time!

- **Grape Skins.** A lot of people want to believe that they'll get massive brain benefits from drinking red wine. I do, too. I'm really sorry to break it to you, but the truth is that the toxic effects of the alcohol, sugar, and, often, mitochondrial toxins from fungus in wine far outweigh the benefits of its small amount of polyphenols, which come from grape skins in the form of resveratrol.

Resveratrol has been studied extensively, and we know that it improves mitochondrial function.[25] In animals, it has been shown to cause a significant increase in aerobic capacity as well as new mitochondrial growth. It has also been shown to protect against obesity and insulin resistance caused by a poor diet.[26]

You can get resveratrol not only from grapes but also from pistachios, blueberries, cranberries, and chocolate. However, the amount of resveratrol fed to animals in the studies above is higher than you'll ever be able to get from food alone. That's why I lay off the wine, eat a variety of resveratrol-rich foods prepared in a way that maximizes polyphenol absorption, and take a supplement, too.[27]

NEUROTRANSMITTER PRECURSOR FOODS

Neurotransmitters are the chemical messengers that send signals from one neuron to another. There are at least one hundred unique types of neurotransmitters in your body, and each has its own specific function. Everything you do requires your neurons to communicate with each other. For this to happen as efficiently as possible, your neurotransmitters must function well.

So how does your body make neurotransmitters? Some are the natural by-products of cellular functions, but most of them are produced in the gut and in the axons of brain cells. In order to make neurotransmitters, your body relies on specific nutrients. If you don't ingest enough of these nutrients, your neurons aren't able to communicate with each other effectively and may even miscommunicate by sending the wrong signals. This can lead to a host of diseases and a marked decline in your performance. Below is a list of the neurotransmitters that most impact your performance and the dietary sources your body needs to make them.

- **Dopamine.** This inhibitory neurotransmitter (meaning it makes your neurons less likely to fire) is best known for its association with reward circuitry in the brain. Cocaine, opium, heroin, and alcohol all increase levels of dopamine—but there are much better ways to boost your dopamine! Too little dopamine is associated with everything from Parkinson's disease to social anxiety, and having sufficient levels of dopamine has a positive impact on your decision making and performance.[28] It is your "motivation molecule" that gives you the willpower to resist impulses and achieve your goals.

  Dopamine is made from protein building blocks called amino acids, in this case the amino acids L-tyrosine and L-phenylalanine. There are studies showing that sunlight exposure (or a tanning lamp) increases dopamine, too!

  Foods high in L-tyrosine:

  - Beef
  - Chicken
  - Turkey
  - Avocados
  - Almonds

  Foods high in L-phenylalanine:

  - Wild salmon
  - Sardines
  - Bacon
  - Beef
  - Liver
  - Almonds

  Of course, you can supplement with these amino acids affordably, as well, but it's hard to get dopamine in supplement form. If you feel that you're experiencing decreased motivation, procrastination, forgetfulness, and mood swings, it's worth trying a tyrosine supplement.

- **Norepinephrine (noradrenaline).** Norepinephrine is an excitatory neurotransmitter (meaning it makes your neurons more likely to fire) that your Labrador brain needs in order to jump into action. It is also key to your performance because it helps you form new memories.

  Dopamine is a precursor to norepinephrine, so the foods listed above can boost your norepinephrine levels, as well. The conversion process from dopamine to norepinephrine requires ascorbic acid (vitamin C), so it's important to eat plenty of green veggies or take a vitamin C supplement daily.

  When I first started biohacking, I wasn't yet thirty and lab tests from my physician told me that I was at high risk of heart attack and stroke. When I measured my neurotransmitter levels, my norepinephrine was seven times higher than it should have been, and four times the level that causes burnout. No wonder I was stressed! I needed other neurotransmitters to balance out my brain.

- **Serotonin.** This inhibitory neurotransmitter has a direct impact on your mood. Having too little serotonin has been linked to depression, anger, and even suicide. As you may already know, many antidepressants work by preventing your neurons from absorbing serotonin so that there is more of it left floating in your synapses. Serotonin also impacts your perception. Hallucinogens attach themselves to serotonin receptor sites and mess with your perception. Proper serotonin levels are also crucial for quality sleep, because your sleep hormone, melatonin, is produced from serotonin.

- **L-tryptophan** is the precursor your body needs to create serotonin. Perhaps you've heard of the "turkey coma," which allegedly comes from eating too much turkey on Thanksgiving. The popular belief is that turkey makes us tired because it contains tryptophan. But the truth is that there isn't enough tryptophan in turkey to knock you out—it's the junk food and massive sugar you eat on Thanksgiving that does that!

Foods high in tryptophan include:

- Lamb
- Beef
- Chicken
- Turkey
- Wild salmon
- Mackerel
- Cashews
- Almonds
- Hazelnuts

- **Acetylcholine.** Acetylcholine was the first neurotransmitter ever discovered and is one of the most important. It stimulates muscles, plays an important role in REM (rapid eye movement) sleep, and is extremely active in the specific pathway in the

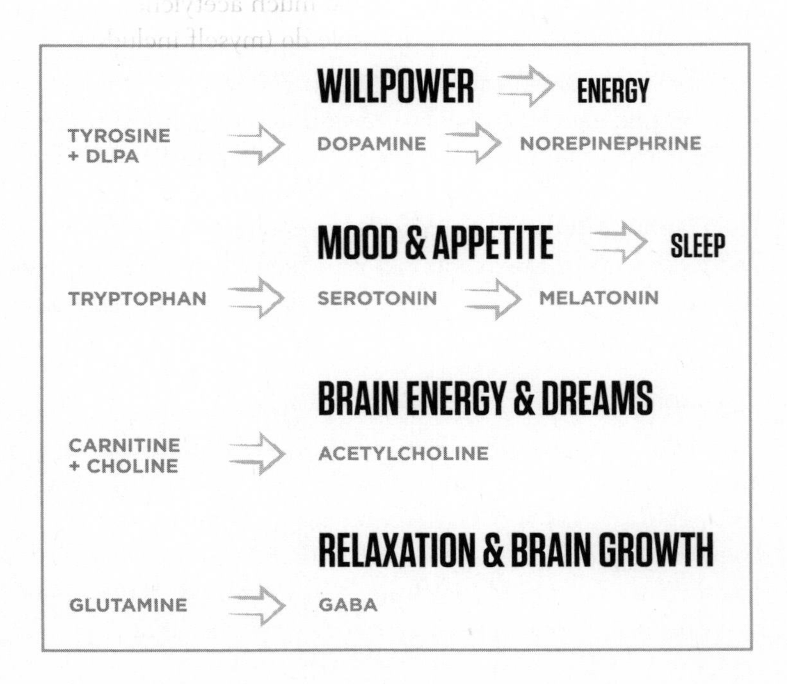

brain that needs to be upheld in order to avoid Alzheimer's disease. Degradation of this pathway is associated with Alzheimer's, and studies show that people suffering from Alzheimer's experience a significant decline in acetylcholine.[29]

The cells in the part of your brain where acetylcholine is so active are the ones responsible for your body's temperature and sleep rhythms. Sleep and temperature regulation are closely connected thanks to acetylcholine. In one study, lab animals that were kept awake lost their ability to regulate their body temperature.[30]

When acetylcholine is low, you're more likely to be tired during the day, and you won't dream as much at night because you won't spend much time in the deepest stage of sleep (REM). I've been boosting my acetylcholine levels with supplements and raw egg yolks for almost twenty years. When I'm tired and stressed, boosting my acetylcholine really makes a difference.

On the other hand, if you have too much acetylcholine, as an estimated 20–30 percent of people do (myself included), adding more through supplements can cause jaw tension, muscle cramps, and teeth grinding at night. You're unlikely to experience problems from eating foods high in acetylcholine, however.

Acetylcholine precursors include L-carnitine and choline, which are best absorbed by the body when consumed together.

Foods high in L-carnitine:

- Beef
- Lamb
- Pork

Foods high in choline:

- Egg yolks (by far the best source)
- Beef

- Kidney
- Liver
- Wild salmon

- **Gamma-Aminobutyric Acid (GABA).** GABA, an inhibitory neurotransmitter, plays many important roles in your brain. It influences the development of new neurons, helps them become differentiated, and forms synapses. But it is perhaps best known for its ability to calm the brain and reduce anxiety. By making your neurons less likely to fire, GABA quiets, or calms, the entire nervous system. Without GABA, your neurons can fire too often and too easily. GABA regulates this activity, and you need enough of it to stay calm in stressful situations. You can think of GABA as a nice, soothing belly rub for your inner Labrador. Anxiety disorders such as panic attacks and seizure disorders are related to low GABA activity. In fact, anxiety medications such as Valium work by enhancing the effects of GABA in the brain.

  Years ago, I had a boss who became incredibly stressed out as our company ballooned in growth from three hundred to five thousand employees in just three years. After I watched her endure a particularly frazzled day, I offered her a GABA supplement. It dramatically changed her performance at work. Years later, I gave two capsules of GABA to a stressed hedge fund manager in Hong Kong before he took off on a flight for London. He called me later and told me that not only had he slept better on the plane than ever before, but he wanted to know if it was also safe to use GABA to stay calm and focused during his high-stakes business transactions. He felt so good that he was concerned about addiction. (Fortunately, not an issue!)

  The precursor to GABA is L-glutamine. Foods high in glutamine include:

  - Beef
  - Lamb
  - Chicken

- Turkey
- Wild salmon
- Eggs
- Organ meats

You've probably noticed some overlap in the foods recommended for the production of various neurotransmitters. It's no coincidence; these are the healthiest and the most beneficial foods you can eat, and you'll get plenty of them on the Head Strong program.

A FAT BRAIN IS A SMART BRAIN

When my daughter was two years old, she sat on Santa's lap at the mall for the first time. He asked her what she wanted for Christmas. Her response? "My own stick of grass-fed butter." I swear I am not making this up.

On Christmas morning, she unwrapped a stick of grass-fed butter, shrieked with joy, and ran around the house holding it above her head like an Olympic torch. No, she didn't eat the whole stick. But she did take one huge, gleeful bite, as if it were a candy bar. Some people accuse me of being a bit too evangelical about grass-fed butter, but I'm so pleased that by the age of two, my daughter already knew that she needed quality fats to help her grow a strong brain (and body).

Over the last few decades, doctors and nutritionists have routinely vilified fat. Beginning in the 1960s we were told that low-fat foods were the "healthy" choice, in large part thanks to the Sugar Association paying a $50,000 bribe to Harvard scientists to encourage them to downplay the risks of sugar and to vilify fat. And it worked. Food manufacturers responded to this messaging by adding sugar to low-fat foods, and we were left with a health crisis and obesity epidemic as a result. Thankfully, more and more people are beginning to reject the phony science that fueled the low-fat craze and embrace the fact that healthy, whole fats are essential. And when it comes to your brain, the right fats are the most important macronutrient there is.

A diet high in healthy fats helps to lower inflammation throughout your body and speed up energy production in your brain. The more healthy fats you eat, the more efficient your brain becomes at converting them to energy. Compared to other macronutrients such as protein and carbohydrates, fat also has the lowest impact on insulin and cortisol levels.

Eating fat is like having your cake and eating it, too. Except you don't get the cake—just the creamy butter frosting!

Our bodies are literally made up of fat. A healthy female body is comprised of 25–29 percent fat, while men's are 15–20 percent fat. Portions of every part of our bodies are made of fat, but none more so than our brains. As you read earlier, our brain cells and the myelin that insulates them are all made of fat and require plenty of high-quality fats to perform at their peak.

When people hear that they should eat more fat, they often worry that doing so will make them have high cholesterol. But like our misunderstanding about the role of fat in a healthy diet, our fears of cholesterol are also misplaced. Cholesterol is not the enemy! In fact, undamaged cholesterol is so vital to the outer coating of your cells that your body actually manufactures it itself. High-density lipoprotein (HDL, a.k.a. "good cholesterol") is necessary and beneficial to your whole body. It removes low-density lipoprotein (LDL, or "bad cholesterol") from the bloodstream and helps maintain the inner walls of blood vessels. But cholesterol is even more important for the brain than it is for your blood vessels. Your oligodendroglia cells (remember them?—they're the cells that maintain myelin) synthesize brain cholesterol and use it to repair and maintain healthy myelin. In fact, 70–80 percent of the cholesterol in an adult brain is found in the myelin sheaths. Low HDL levels are associated with cognitive impairment and neurodegenerative disease,[31] most likely because the oligodendroglia do not have the materials they need to keep your myelin intact.

Of course, not all fats are created equal. Nutritionist and trans-fat researcher Mary Enig, PhD, has explained two helpful ways to understand fats. The first is to look at the length of a fat molecule. As a general rule, the shorter the fat, the more anti-inflammatory properties it offers. Short- and medium-chain fats, which include the butyric acid found in butter and two of the four types of medium-chain triglycerides (MCTs) found in coconut oil, are important choices. In addition to being anti-inflammatory, butyric acid also helps to protect the blood-brain barrier. And eating the butyric acid found in delicious, creamy butter has beneficial effects different from the ones you get when your gut bacteria make butyric acid from fiber. If that isn't a case for the inherent wisdom in nature, I don't know what is.

The second way to understand the value of a fat is to look at how stable it is. Oxygen, as you now know, drives chemical reactions in the body. These chemical reactions can damage fats, some more easily than others. Oxidized (damaged) fats create inflammation in the body. When your body uses these damaged fats to make cell membranes, they are less flexible and less effective, and they create harmful free radicals. These free radicals cause even more inflammation and age you quickly. Damaged fats are the enemy of performance, starting in your brain.

Your body wisely uses many of the most stable fats to which it has access when making cell membranes and hormones. These are saturated fats, which have the fewest places for oxygen to cause damage through oxidation, and just enough of the damage-prone omega-3 fats to make membranes work. When you eat saturated fats, your brain can create strong, stable cell membranes. The second most stable fats are monounsaturated fats, which have only one vulnerable spot where oxygen can get in and wreak havoc ("mono" means one). Unsaturated fats are the least stable and the most inflammatory type of fats, but the brain does need some of them. Omega-3s and omega-6s are both essential unsaturated fats.

As you may recall from chapter 4, omega-3 fats are anti-inflammatory, while excessive omega-6 fats are inflammatory; your brain needs the right ratio of these fats for optimal function. But omega-3s offer more than just anti-inflammatory benefits. You may already be familiar with the two most important types of omega-3 fatty acids for your brain: eicosapentaenoic acid (EPA) and docosahexaenoic acid (DHA). EPA is a powerful anti-inflammatory, while DHA is the primary structural fatty acid in your brain, your retina, and your central nervous system and is essential for brain development.[32]

Go back and read that last sentence again. DHA is the *primary structural fatty acid in the human brain*. In fact, some researchers believe that it was an increase in our intake of DHA that allowed humans to grow such large and powerful brains.[33] Yet, we cannot make DHA ourselves—we have to get it from dietary sources. It is in great supply in human breast milk because it is essential for a baby's growing brain. Even before a woman is pregnant, her body stores extra DHA in her hips and rear end. Some research suggests that men have evolved to prefer curvy body types because those curves signal a woman's ability to have healthy children. By the time a woman gives birth to her second child, her supply of DHA has often dwindled. Some theorize that this is why, on average, firstborn children score higher on aptitude tests than their younger siblings.

You need plenty of DHA if you are literally growing a human brain in your uterus, but DHA is just as necessary for you to maintain a highly performing adult human brain—whether or not you are pregnant. A widespread study of 485 people age fifty-five and over showed that those who took DHA supplements for six months demonstrated significant improvements in memory and learning. The study also showed that higher DHA intake is inversely correlated with the relative risk of developing Alzheimer's disease.[34] In other words, the more DHA you consume, the less likely you are to get Alzheimer's. Studies on mice have shown that DHA improves memory and slows the progression of Alzheimer's disease,[35] and we have every reason to believe the same is true in humans.

Here are some of the best sources of fat to maximize your smart, sexy, fatty brain:

Best Sources of Stable Saturated Fats

- **Grass-Fed Animal Fat and Meat** (bone marrow, lard, etc., but not poultry fat). A 2006 study[36] showed that grass-fed cows have more healthy omega-3s and more conjugated linoleic acid (CLA), which is a type of naturally occurring trans-fatty acid that improves brain function, causes weight loss, and reduces your risk of cancer. Just eighty days of grain feeding was enough to destroy the omega-3 and CLA content of the beef, but the longer the animal ate grains, the lower the quality of the meat. The omega-3 quantity in grain-fed meat was so low that it didn't qualify as a meaningful dietary source, while the grass-fed meat had enough omega-3 to be considered a good source of these fats. In 2008, another study[37] showed that grass-fed meat had slightly less total fat than grain-fed meat, but the real difference was the type of fat. Grass-fed meat was higher in saturated fat, omega-3s, CLA, and trans-vaccenic acid (TVA), which is similar to CLA. The grain-fed and grass-fed animals had about the same amount of omega-6, total polyunsaturated fat, and cholesterol. This means the grass-fed meat had a better ratio of omega-6 to omega-3 fatty acids and healthier fats overall. Bone marrow is especially high in omega-3 fats, and some researchers believe that the practice of cracking open scavenged bones and consuming the marrow was what allowed our ancestors to develop large brains.

- **Grass-Fed Beef Tallow.** Tallow is sort of like butter that's made of animal fat instead of milk fat. It's solid at room temperature, which means it's extremely stable. Grass-fed tallow also has a good omega-6 to omega-3 ratio and can be delicious and good for you, especially if it's prepared with care.

- **Pastured Egg Yolks.** Just like butter, egg yolks lose most of their nutritional benefits when they come from an animal that was fed antibiotics and GMO (genetically modified organism) corn and soy, which is the diet of essentially all chickens you'll find in supermarkets. Even organic chickens eat mostly organic grains. Sure, that's better than eating GMO grains, but not by much. Pastured egg yolks, from chickens that eat grass and have the freedom to roam wherever they like, have a deep golden color from their vitamin A and antioxidants and contain twice as many omega-3s as "regular" eggs.

  Hold on, I want to say that again—*twice as many.*

  You'll want to eat your yolks runny, as in sunny-side up, soft-boiled, or poached, so you don't damage the valuable cholesterol and phospholipids found in yolks. Hard-cooked eggs are much more inflammatory.

Best Source of Monounsaturated Fat

- **Olive Oil.** Olive oil contains oleocanthal and oleuropein, two anti-inflammatory and highly potent antioxidants. Oleocanthal is almost medicinal in its health effects. It has been shown in studies to clear the brain of the dangerous amyloid plaques that are linked to Alzheimer's, and it also causes the death of cancer cells. However, it's important to treat olive oil with care because as a monounsaturated fat, it is less stable than proper cooking fats such as ghee and coconut oil. Oleocanthal is easily damaged when heated, so it's best not to heat or cook with extra-virgin olive oil. Just add it to salads or cooked foods after they've been heated. Olive oil can oxidize when exposed to light, so look for olive oil that comes bottled in dark glass.

  Recent investigations into the olive oil industry have revealed that a lot of companies illegally dilute their olive oil with cheaper oils, so it's important to find a brand you trust. California olive oils are a good choice because they're usually produced by smaller companies and are often fresher than European options. Olive oil is heavy and expensive to ship by air, so most European companies send it to the United States by boat. By the time you take the olive oil off the shelf, it's often several months old, and some of its beneficial compounds have degraded.

Best Sources of Omega-3 Fatty Acids

- **Wild-Caught, Low-Mercury Seafood.** Wild-caught seafood is high in healthy fats, micronutrients, micro-minerals, and antioxidants. Farmed seafood can be high in pesticides, toxins, heavy metals, parasites, pathogens, and environmental contaminants. It's also much lower in nutrients and healthy fats than wild-caught seafood. All fish contain some mercury, and your brain can handle small amounts, but it's best to choose seafood with the lowest possible levels.

Of course different types of seafood also offer varying amounts of omega-3s. Many fish are higher in DHA than EPA, yet all contain some of both fatty acids. The following choices are relatively high in omega-3s and low in mercury:

- Sardines
- Sockeye salmon
- Anchovies
- Mackerel
- Wild trout

- **Fish/Krill Oil.** Many people buy fish oil to increase their omega-3 intake, but poor-quality fish oil can cause more problems than it solves. Many of the discount brands you are likely to see at your local grocery store are contaminated, oxidized, and low potency. They smell extra fishy when you open a capsule. If you can't find a good fish oil, you're much better off avoiding it altogether.

  This is why I recommend krill oil over fish oil. Krill oil is more stable and is phosphorylated, meaning it's easier for your brain to use. It also contains astaxanthin, a potent antioxidant shown to improve mitochondria.

If it's not clear by now, it bears repeating: to help your brain work at its fullest potential, you're going to consume the right kind of fat. It is the most important macronutrient for your brain, and, as a bonus, it's satiating and tastes good. Fat kills cravings. You'll never feel hungry or deprived while you're eating the healthy fats on the Head Strong program.

KETOSIS IS THE MOSTEST

In the year 400 BC, the Greek physician who later came to be known as the "father of Western medicine," Hippocrates, wrote about a miraculous recovery he had witnessed. When a man who suffered from debilitating seizures abstained from food, he experienced an incredible improvement in his symptoms. At the time, of course, there was no name for epilepsy or an understanding of the seizures it caused. But it was clear that when a person who was suffering stopped eating, the seizures also stopped.

It wasn't until the twentieth century that the scientist George Cahill learned that fasting prevents seizures by putting the patient into a state known as ketosis. During periods of fasting or severe carbohydrate restriction, the liver breaks down fatty acids to produce ketone bodies, water-soluble molecules that are the ideal fuel for our mitochondria, much better than sugar. When your mitochondria are using ketones as fuel to create ATP, you are in a state of ketosis. This is a state of high performance (but one you might not want to be in all of the time). It increases your mitochondria's energetic output, reduces your production of free radicals, and increases your production of the important inhibitory neurotransmitter GABA. Later on, I'll explain the different ways to get into ketosis.

It turns out that ketosis protects epileptics from that double-edged sword, oxygen. Your cells need oxygen to make energy, but oxygen stimulates neurons, and too much neuron stimulation causes excitotoxicity. This is a process that kills neurons by over-activating neurotransmitter receptors and generating oxidative stress and can cause seizures. Ketosis increases GABA and antioxidant levels, which helps prevent seizures. It doesn't hurt that your mitochondria create ATP more efficiently from ketones than glucose, creating less oxidative stress in the process[38] and providing more energy for your neurons to control excitotoxcity. And as you've already learned, when your mitochondria are working better, they make fewer free radicals.

It seems that ketosis is meant to protect us from all sorts of threats. If it weren't for ketosis, we would starve in times of scarcity because our mitochondria would not be able to produce ATP from our fat.

Since we've only had access to food at any time of day for the last few hundred years or so, it's pretty obvious that ketones have allowed us to survive this long as a species. When food was less abundant, ketones provided resilience against hypoglycemia and, ultimately, starvation.

What makes ketosis so efficient? As you may recall from chapter 2, the first step to creating energy is to get fuel inside your mitochondria. Unlike glucose, ketone bodies can get into the mitochondria completely and intact. They are then broken down directly into acetyl CoA, which enters the Krebs cycle. Glucose, on the other hand, must first get broken down outside the mitochondria into pyruvate. Then pyruvate is broken down into acetyl CoA inside the mitochondria. Long story short: ketones are able to skip a step before they even get to work creating energy.

This increase of acetyl CoA from ketones also drives the Krebs cycle to "charge up" more NADH, the molecule that provides energy to the electron transport system. Simple version: ketosis makes your body's process of releasing and recycling energy go faster. This uses up less oxygen than burning glucose and creates less oxidative stress in the process.

And because ketone bodies are more energy-rich than pyruvate, they produce more ATP. While 100 grams of glucose generates 8.7 kilograms of ATP, 100 grams of a ketone body can yield 10.5 kilograms of ATP.[39] That's over 20 percent more!

Remember, most of your cells have several hundred mitochondria each. The cells in your brain each have thousands. So an increase in energy production by even as little as a couple of kilograms of ATP is huge. It can be the difference between just slogging by and feeling like you're on fire (in a good way).

I recently had the great pleasure of interviewing Dr. Richard Veech, a renowned ketone expert who studied with Hans Krebs, the very man who discovered (and named) the Krebs cycle. Dr. Veech explained that humans are the only animals that go into ketosis when we fast because we have such large brains to support. Ketosis protects our big brains from oxidative stress and allows us to survive. Without ketones, we would die in six days without food, but with them we can survive much longer. Before we had McDonald's on every corner, we fasted in between periods of hunting and killing animals. And during those times, we thrived in a state of ketosis.

Personally, I know that being in ketosis makes a huge impact on my energy level, but I was never able to put an exact number on just how much more energized I felt in this state. In our interview, Dr. Veech told me that the heart gets 28 percent more energy when it is metabolizing ketone bodies compared to glucose. Since the brain and the heart have roughly the same density of mitochondria, ketones' impact on the brain is just as dramatic.

This number blew my mind. Can you imagine what it would be like to have 28 percent more brain energy? I can't emphasize enough how important this is. In my twenty years of seeking new ways to improve my performance, I haven't found anything more impactful than ketones. I've found things that moved the needle 5 percent or 10 percent, but nowhere near the 28 percent of ketones. Think of it this way: if ATP provides the energy for you to be you, being in ketosis allows you to be 28 percent more you. The only question is: what will you use all that extra brainpower to accomplish?

Over time, your body gets more and more efficient at running on ketones. When you do, you become what's known as "keto adapted." Being keto adapted has its benefits. Your body becomes more efficient at breaking down fats and burning them for energy. You also experience powerful detoxification effects. Fair warning: this might not feel so good in the beginning. In fact, the first time you go into ketosis, you're going to get the worst breath ever. This is because your body stores a lot of toxins in fat. When you start burning that fat to create energy, your liver must process the toxins that are released. For a few days, you might feel tired and foggy, unless you take toxin binders such as activated coconut charcoal. Otherwise, as your liver processes all of the toxins that have been released from your fat, it pulls energy away from your brain.

This adjustment period is short and more than worth it to experience what follows. Once your system is running clean, a new level of energy and performance will replace your bad breath, fatigue, and brain fog. I've heard people say that they felt truly alive for the first time after experiencing ketosis, and I don't think that's an exaggeration at all.

Of course, nothing is perfect, not even ketosis. I've found that any time you get your body into a certain state and try to hold it there forever, it starts to resist. The two most common ways of getting into ketosis are to almost entirely restrict carbs or fast completely. If you maintain a ketogenic (extremely low carb) diet for months on end, your cortisol levels may rise and start tearing up muscle mass to get glucose for your brain. Even when fasting, your blood glucose levels will never go to zero because your body will keep increasing cortisol production. Cortisol makes cells insulin resistant so they cannot synthesize glucose. This causes blood glucose levels to remain high, awakens your inner Labrador, and leads to major cravings.

This explains why many people (but not all) start to feel lousy after several months in continuous, uninterrupted ketosis. It was certainly the case for me. When I was fat, I lost fifty pounds on an Atkins-type diet and felt amazing—until suddenly I didn't anymore. Besides cravings, one of the first symptoms of cutting too many carbs for too long is having very dry eyes. Being this low in carbs without a break can also ruin your quality of sleep. Eating no carbs for extended periods can also damage your thyroid,[40] which is a big problem since you need thyroid hormones to make ATP and maintain healthy myelin. (That said, some people seem to thrive in continuous ketosis—though they are in the minority.)

I also learned that while neurons prefer ketones as an energy source, other cells in the body, such as the ones that repair and maintain myelin, actually prefer glucose. Your body is like a hybrid car that runs most efficiently when it has the option of alternating between fuel sources. This inspired me to find a way to hack ketosis. How could I enjoy the extra energy it gave my brain without depriving my other cells of glucose and without experiencing the negative side effects?

I found that there are ways to experience mild ketosis without fully restricting carbohydrates. In the 1980s, researchers at Harvard Medical School discovered how to produce medium-chain triglycerides (MCTs) in large quantities and use them for therapeutic purposes,[41] although the high doses required to raise ketones often led to a very unpleasant gastrointestinal state that I affectionately call "disaster pants." They learned that because some MCTs are water-soluble, they go directly to the liver, and your liver metabolizes them on the spot.

To do so, it wipes out your storage of glycogen. This puts you in a state of ketosis very quickly, even if some carbohydrates are present.

This was a game-changing discovery for me, which led me to start to use MCT oils, despite the gastric problems. And I realized very quickly that it was worth a few extra trips to the bathroom to feel this good! Later, I developed my own Brain Octane Oil, which—according to a study from the University of California, San Diego—works significantly better than MCT oil to raise ketones in the blood. Brain Octane Oil is designed to amplify ketosis and give you the best of both worlds. When your blood concentration of ketones is 0.8 millimolar, you are considered to be in a state of nutritional ketosis. But I've found research that says that having even a small amount of ketones in your bloodstream—a mere 0.5 millimolar—is enough to shift two hormones that control hunger and fulfillment.[42] I can reliably reach this level of mild ketosis with Brain Octane Oil even when eating some carbs. All of a sudden, my cravings disappear and my focus and concentration go way up. And disaster pants is much less of an issue!

There is a well-known way to reach a state of ketosis without Brain Octane Oil. The trick is that you have to reduce your carbs to about 20 grams per day and eat 70–85 percent of your calories from fat. At that level, you have to cut veggies and their valuable polyphenols, and over time you have to be careful to avoid nutrient deficiencies. With Brain Octane Oil, you have a buffer, and you can get the best of both worlds—you can eat some carbs every day but still have some ketones in your blood.

FASTING FOR A FAST-ACTING BRAIN

Of course, restricting carbs and consuming Brain Octane Oil aren't the only ways to get into ketosis. There's also good, old-fashioned fasting. Fasting has some additional benefits beyond putting you into a ketogenic state. Fasting improves myelination and myelin regeneration in the brain and helps reduce inflammation throughout the body. Fasting also switches on your cellular detox system, autophagy. When this happens, you eliminate cellular waste. It is the equivalent of a deep cleaning for your brain, and it feels amazing.

But nothing is perfect. Fasting has some drawbacks, like making you really pissed off. That's because when you stop eating, the brain thinks it's dying. So all of your most primitive emotions—rage, anger, and sadness—rise to the surface. This is not fun for anyone, especially your coworkers and family!

Lots of people have tried to hack long-term fasting with a technique called intermittent fasting. This means you eat all of your food within a specific window of time (six or eight hours) and fast the rest of the day (sixteen to eighteen hours). This is an effective way to cycle in and out of ketosis while remaining well fed, and it's been proven to have plenty of other brain benefits such as increasing neurogenesis.[43]

The downside to intermittent fasting is that it causes energy slumps. Most intermittent fasting protocols require you to skip breakfast and not eat lunch until after two p.m. When I tried this, I found that my brain energy took a deep dive in the middle of my workday, right when I really needed to bring the brainpower. I wondered if there was a way to get the benefits of ketosis and intermittent fasting without crashing.

I had already created Bulletproof Coffee at this point, and I knew how it made me feel, if not all of the reasons why. It was inspired by the polyphenol-rich yak butter tea that Tibetans drink in the morning before meditation, and it evolved to become a careful blend of Brain Octane Oil, grass-fed butter, and coffee brewed with special-process beans free of mitochondria-inhibiting toxins. Coffee is a great source of polyphenols, but it often has mold toxins that slow down your mitochondria. So I created a new process to make coffee without the risks of mold toxins like normal coffee and decided to combine it with grass-fed butter and Brain Octane Oil. Grass-fed butter gave me the anti-inflammatory butyrate, and Brain Octane Oil helped put me in a mild state of ketosis. It turns out that this combination gave me what my brain needed most, and without any protein or sugar, so that my body thought I was fasting. I tried Bulletproof Coffee in the morning during an intermittent fast, and voilà: I felt all of the benefits of the fast but with much more energy than ever before.

When I drink a cup of Bulletproof Coffee in the morning instead of eating breakfast, my body remains in fasting mode, yet I feel satiated and energized. Hundreds of thousands of people have tried it now, and many feel the same thing. The Brain Octane Oil puts me into a mild form of ketosis even if I ate some carbs the night before. The polyphenols in coffee fight inflammation, support new brain cells, and feed more good bacteria in my gut. And the butyrate in the grass-fed butter protects my blood-brain barrier while continuing to feed the good gut bacteria. It also helps make the polyphenols from the coffee more bioavailable, and it doesn't contain the milk protein that locks polyphenols away from my brain. It is safe to say that this has been the most life-changing nutritional performance hack I have ever experienced.

The best part is that the fat from the butter and the Brain Octane Oil shuts off my food cravings as quickly as flipping a switch. With my inner Labrador calmly resting at my feet, I am able to focus for hours at a time with a new level of intensity that powers me through my day. I am not the least bit interested in food, but my cells still believe I am fasting. I believe that there is truly no better, faster, or more effective way to become Head Strong.

If you don't believe me, all you have to do is try it one time—coffee made with the right beans, Brain Octane Oil, and grass-fed butter. The difference is like night and day. The truth is that you can make plain butter coffee with any old beans, any old butter, and coconut or MCT oil—but the science says you'll also get mitochondrial-inhibiting toxins, some of the wrong fats, and far fewer ketones, and experience says you won't feel as Head Strong as you could. The benefits of Bulletproof Coffee will go a long way toward optimizing your brain.

Head Points: Don't Forget These Three Things

- Polyphenols are antioxidants that protect your gut, increase your rate of neurogenesis, play a role in apoptosis, and lower inflammation—and your body needs fat to absorb them.
- You need certain nutrients to make the most important neurotransmitters for your brain. The best foods for this include beef, almonds, eggs, lamb, and wild salmon.

- Your mitochondria make energy more efficiently with ketones than they do with glucose.

Head Start: Do These Three Things Right Now

- Always eat low-sugar fruits and veggies with a healthy fat source such as grass-fed butter.
- Eat more fish to get plenty of essential fatty acids for your brain.
- Use Brain Octane Oil or restrict carbohydrates so your brain has access to ketones.

6

BRAIN-INHIBITING FOODS

You know that the foods you eat can help to fuel your brain and increase your performance—which is great news, because it means that the tools to building a supercharged brain are well within your reach.

But just as your food choices can enhance your mental power, they can also hurt it. Some of the foods you eat every day are actually damaging your neurons, making your mitochondria less efficient, slowing your cellular energy production, fueling inflammation, and making you feel cranky, distracted, forgetful, and foggy—sometimes all at once. And you probably don't even know it's happening.

The whole point of eating food is to provide our bodies with the energy they need to fuel all of our bodily systems, especially the organ that consumes the most energy—the brain. We're learning more every day about the biology of energy production, but one thing we know for sure right now is that certain foods actually rob your body of energy. And we can use that knowledge to start feeling a lot more Head Strong today.

INFLAMMATORY FOODS

As we discussed in chapter 4, some foods cause inflammation by irritating the lining of the gut and triggering your immune system to attack healthy cells. These foods will always have a negative impact on your brain. When the mitochondria are inflamed, the electron transport chain becomes elongated, and it takes more time for electrons to move through the chain. The result is that you become less efficient at producing energy, your performance suffers, and you have less of the energy it takes to be yourself. It's as simple as that.

Unfortunately, it's not always easy to know which foods are causing your inflammation. That's because inflammation manifests quietly—you might start to feel the symptoms once your body has become highly inflamed, but most of the time you'll just notice a muffin top forming around your middle a day after you ate a mitochondria-inhibiting food and figure it's because you haven't gone to the gym in a while. What you can't see is the muffin top of inflammation forming in your brain. Instead, you'll experience symptoms that seem completely unrelated to your doughy midsection. You might become easily irritated and get snippy with the people you care about the most. I've said some of the worst things to my family when food or toxins disrupted my mitochondria and made the emotional processing parts of my brain ineffective. I'm a peaceful guy . . . except when my energy production systems are poisoned. Then I'm kind of a jerk.

To make things even more complicated, some people's immune systems are more sensitive to certain foods than others. You may not even realize that you have sensitivities to some of the foods you eat on a regular basis. This is why I recommend seeing your doctor for a food allergy blood panel. This simple test can show you exactly which foods may be an issue for you.

While a blood test is the gold standard, there is a free option. It is less accurate, but it's still a helpful hack. When you eat a food you're sensitive to, your heart rate increases by about seventeen beats per minute within ninety minutes of eating your food. So if you're going to eat something that you think might be causing a problem, take your heart rate before a meal and several times after, looking for a big increase. If your heart rate is substantially higher, you may have found a guilty suspect! I have a free smartphone app called Bulletproof Food Detective that helps you with this test.

In the meantime, there are some foods that cause inflammation in nearly everyone. You should avoid these foods as much as possible, because they always cause problems. You may not feel the impact until a day after an exposure, so it's often hard to connect cause with effect. Here are the top offenders:

- **Trans Fats.** Trans fats are born when hydrogen is added to liquid vegetable oils to make them more stable. Food manufacturers like to use these fats in their products because they have a longer shelf life than other fats. Trans fats also form in restaurant fryers over time. Unfortunately, trans fats wreak havoc in your body and are your brain's nemesis. Trans fats change the composition of mitochondria[1] and build up inside the mitochondrial matrix as they are metabolized.[2] They also lead to inflammation in your brain and cause your immune system to become overactive.[3] Consumption of trans fats has been linked to a host of illnesses including cancer, dementia, Alzheimer's disease, liver damage, infertility, and depression. Of course, those are all mitochondrial disorders!

  Many countries have begun to crack down on trans fats, but they're still found in tons of products, particularly baked goods, fried foods, nondairy creamers, potato chips, and margarine. And they're abundant in fried foods from restaurants, even though they're not listed on the label of the fryer oil. These are all foods that you should avoid like the plague. Your mitochondria may not work properly for several days after you eat trans fats, so it's not okay to "cheat" on this one.

- **Dairy Products.** Milk is basically made up of three components (besides water): protein, sugar, and fat. When people are sensitive to dairy, it's usually the sugars or proteins that are affecting them. You've probably heard of lactose, the main sugar in dairy products. If you are lactose intolerant, you are missing lactase, the enzyme that your body needs to digest lactose. If you are lactose intolerant and still choose to consume dairy products, you'll experience systemic inflammation and weakness every time you have a glass of milk or a scoop of ice cream.

  But even people who are not lactose intolerant are often sensitive to the main protein in dairy products, casein. Casein has a similar molecular structure to gluten, and people who are sensitive to one are often also sensitive to the other. In many people, casein breaks down into casomorphin, which binds to the opiate receptors in your brain and makes you feel sedated.

Even worse, milk protein isolates are common ingredients in popular "keto" or "low-carb" protein bars. This is not what you want in a highly performing brain!

Milk protein causes inflammation, which lowers mitochondrial function. Milk protein also binds to the good-for-you polyphenols in your diet (like the ones found in your coffee!) and makes them unavailable to your body—and your mitochondria. When you put milk protein isolates and casein into your coffee by adding milk, half-and-half, cream, or other creamers, the polyphenols become 3.4 times less absorbable in your body.[4] The same thing happens if you use black coffee to wash down a milk protein–based "keto protein bar." So not only does milk protein directly change your mitochondria for the worse, it also prevents you from absorbing the nutrients that can repair and strengthen your mitochondria.

I know what you're probably thinking: *Isn't butter made from milk?* Yes, technically butter is a dairy product, except that butter contains almost no milk protein. A tablespoon of grass-fed butter has 0.1 gram of milk protein and almost no lactose. Cultured butter has even less than that, and ghee (clarified butter) has microscopically small amounts. Butter yet (ahem), animal studies show that the short-chain fat found in butter, butyric acid, causes "an increase in mitochondrial function and biogenesis."[5] That's why I use butter instead of milk in Bulletproof Coffee and why it's a part of the Head Strong program. Milk *protein* is bad for your mitochondria. Milk *sugar* is still sugar. But milk *fat*? *That's* what does a body good.

- **Gluten.** Unless you've been living under a rock, you've probably heard that gluten, the protein found in wheat, causes digestive distress for many people. It's estimated that three million Americans have a gluten allergy (celiac disease) and up to eighteen million have gluten sensitivity. But gluten affects us all.

  When you consume foods that contain gluten—such as bread, pasta, and cereal—your body is prompted to release a protein called zonulin, which controls the space between the cells that line your digestive tract.[6] Gluten prompts an over-release of zonulin, which pushes these cells farther apart, allowing pathogens to slip through the protective barrier of your gut lining (leading to a condition known as leaky gut syndrome). The result? You guessed it: inflammation. Gluten also reduces blood flow to the brain and interferes with the thyroid hormones[7] you need to create ATP, maintain healthy myelin, and have healthy mitochondria. This is true for all of us (not just those who are allergic to gluten), and it can lead to problems with learning, focus, and memory.

If you already have mitochondrial dysfunction (recall that 48 percent of us may), gluten is especially problematic. Studies have shown that mitochondrial dysfunction causes leaky gut in animals[8] and is tied to inflammatory bowel disease.[9] Gluten just adds more inflammatory fuel to the fire.

As an accomplished home chef and the author of *Bulletproof: The Cookbook*, I'm sad to be the bearer of this news. Gluten tastes good and you can create so many delicious things with it . . . but the long-term cost is not worth the short-term benefits. Life is about how you feel all of the time, not just how delicious a few bites of food are. So ditch the gluten and see how much better you feel—the longer you avoid it, the more time your body has to get stronger.

- **Vegetable Oils.** Canola, corn, cottonseed, peanut, safflower, soybean, sunflower, and all other vegetable oils are inflammatory. Why? Because the way they are produced damages most of these oils, and they are so unstable that they easily oxidize in heat, light, or air. They also provide an overabundance of omega-6 polyunsaturated fats. Remember, you need a good balance of omega-3s and omega-6s to keep inflammation at bay (the perfect ratio is roughly one omega-3 for every four omega-6s you consume). But because these oils are used in so many packaged and processed foods, most of us who consume a Western diet eat more than *twenty* times more omega-6s than omega-3s. This imbalance is one of the leading causes of inflammation. You'll get the right amount of undamaged omega-6 fats when you follow the Head Strong program.

TOXIC FOOD

There are two types of toxins in your food: the toxins that manufacturers add to food, such as preservatives, pesticides, and artificial flavors, and the naturally occurring toxins that plants, bacteria, and fungi form to protect themselves from being eaten by animals or bugs. These two types of toxins may come from vastly different sources, but both can make your mitochondria weak and cause inflammation, brain fog, and diminished mental performance. Let's take a look at some of the worst ones.

MOLD TOXINS

Mold toxins (also called mycotoxins) occur naturally in many foods and can be found growing in many environments. Whether it's in your food or in your house, mold can negatively impact your brain. Many mold toxins are direct mitochondrial toxins. This is a huge problem, but few people talk about it. As a kid, I lived in a water-damaged basement (of course we didn't know it was moldy at the time), and I was always sick. I had chronic sinus infections, asthma, and even arthritis as a young teen, not to mention obesity and ADD (attention deficit disorder)! It took many years for me to realize that my symptoms, especially the mental and emotional ones, largely stemmed from toxic mold exposure and the resultant damage to my mitochondria. Environmental mold made me Head Mushy, not Head Strong.

To this day, decades after getting my brain (and body) back, I remain sensitive to mold. This is how I first became aware of the dangers of the mold toxins in our food supply. I wouldn't wish the suffering I went through as a kid and young adult on anyone, but now I'm grateful for my experiences with mold because I understand how ingesting even small amounts of mold toxins can hinder mitochondrial performance. Many studies have shown that common mold toxins in food cause mitochondrial dysfunction.[10] But most people don't know when they've been exposed to mold, and they don't realize that an "off" day might just be a sign of mitochondrial dysfunction.

Because of my particular genetics, I fall into the 28 percent of the population that gets extra inflammation from mold. If you have this sensitivity, walking into a musty room or drinking a cup of moldy coffee in the morning can knock you out for the rest of the day. The other 72 percent of people might experience a mild brain fog later in the morning, food cravings at lunchtime, or irritability. For most people, the effects of mold toxins are subtle enough to write them off as "having a bad day," "feeling cranky," or "needing more sleep." You might even consider them "normal," but they're still silently slowing you down. And they're not inherent character flaws.

Because I'm so sensitive to mold toxins, I usually know right away when I've consumed even a small amount of them. I've learned how to avoid them (almost) completely, and along the way I've been lucky enough to help other people eliminate mold from their lives and start performing at their maximum potential, even people who have no idea it's holding them back. I filmed a documentary about mold toxins called *Moldy*, which features a dozen experts covering the massive extent of this problem. You can view the film at moldymovie.com/headstrong2017.

At high levels, mold toxins can cause serious illnesses such as cardiomyopathy, cancer, hypertension, kidney disease, and brain damage. But consuming or being exposed to even small amounts of mold can make you feel sluggish and distracted. Our big energy-hungry brains and our relatively slow detox processes make us the most susceptible mammals on Earth to the effects of mold toxins.

It makes a lot of sense when you think about it. The war between bacteria and fungi has been raging for millennia, long before mammals (let alone humans) came on the scene. Mold is the source of antibiotics—drugs designed to kill bacteria. And, when they can, bacteria will try to kill mold. Hundreds of millions of years of evolution have made fungi really good at surviving the chemical weapons of bacteria. But we—humans—are basically big bags of bacteria. Almost every cell in your body contains thousands of mitochondria, which are bacteria that support your cells. Is it any wonder that fungi, the ancient enemy of bacteria, have the ability to damage the power plants in your cells and wreak havoc in your brain? And that's not even considering what they do to your gut bacteria.

You might be tempted to write this all off and assume you haven't been exposed to mold toxins. But according to the experts in *Moldy*, about half of the buildings in the United States have problems with water damage. And mold toxins are an incredibly common contaminant in our food supply. Governments around the world have finally started paying attention to this little-known problem.

One type of mold toxin, called ochratoxin A (OTA)—found especially in high-polyphenol foods such as coffee, chocolate, wine, grains, and beer—is pure mitochondrial kryptonite. In many studies, OTA has been shown to interfere with apoptosis, impair mitochondria, cause oxidative stress, and make mitochondrial membranes more permeable.[11] OTA also impairs your cells' antioxidant defenses, making them more susceptible to the oxidative stress caused by OTA. This slows down your energy production and accelerates the aging process, all while killing off your healthy cells. OTA is also an immune suppressant and makes you more susceptible to cancer and various autoimmune diseases.[12]

In one study, scientists looked at the effects of OTA on rat liver mitochondria. You might not want to think about rat livers, but the results of the study were fascinating—and scary. After isolating the mitochondria and exposing them to OTA, the researchers found that mitochondrial function declined in proportion to the concentration of OTA used to treat the cells. The OTA slowed down energy production at first; with increased exposure, eventually the entire energy production system broke down.[13] And the scary thing is that rats are far better at eliminating OTA than humans. We are one of the few animals that use our relatively weak kidneys to excrete it. Rats and most other animals use their much stronger livers to excrete it. Despite this difference, OTA still harms rats, but not as much as it harms people.

OTA also causes *permanent* changes to the structure of mitochondria. In another study, mitochondria that were exposed to OTA deteriorated, becoming swollen as their energy production decreased.[14] Yet another disturbing study found that OTA inhibited embryonic development in pregnant mice by overstimulating apoptosis (cell death).[15]

Some scientists now believe that the mitochondrial damage caused by mycotoxins (particularly OTA) may be a leading cause of chronic fatigue syndrome. In one study, doctors tested the urine of patients who had previously been diagnosed with chronic fatigue syndrome. A whopping 93 percent of them tested positive for mycotoxins, compared to 0 percent in the healthy control group. OTA was the most prevalent type of mycotoxin present and was detected in 83 percent of the patients.[16] No, we're not all going to develop chronic fatigue syndrome, but what about mild fatigue, brain fog, trouble remembering things, and a general lack of awesomeness? These symptoms can be caused by OTA exposure, too. There are over nine hundred studies about OTA's effects on humans on the Bulletproof website.

Unfortunately, it's difficult to avoid consuming OTA and other mycotoxins, especially in the United States, where the government does not regulate OTA levels to the extent that many other countries do. Even in Europe, where regulations do exist, I believe the allowable limits are too high. Government "safe" limits on mold toxins are not designed to make us feel amazing—they're designed to protect food manufacturers from losses when their food goes bad too quickly.

OTA is just one mold toxin. There are about two hundred other identified toxins from molds. These toxins amplify each other, so a government-declared "safe" amount of OTA may not be safe at all if you eat it with other common toxins that are not regulated. At the time I'm writing this, there is no government on Earth with multi–mycotoxin exposure regulations, despite clear evidence that they are necessary. That's why I test the products I make for more than twenty common mold toxins.

The good news is that you don't have to *completely* give up the foods that commonly contain mold toxins, but minimizing your consumption of them is a key component of supercharging your brain. So is learning to feel their effects.

Learning the toxic effects of mold can help you understand the difference between an amazing day and an awful day, why you slipped into a food coma after one meal but felt energized after another meal with the same ingredients, or why you felt like crap after spending a day at your friend's house, which had water stains on the ceiling.

Here are some of the most common sources of these mitochondrial poisons:

CEREALS AND GRAINS

A 2016 test of more than eight thousand grain samples found that 96 percent contained at least ten types of mycotoxin,[17] and another study found that at least 25 percent of the world's grain crops are contaminated with OTA.[18] After harvest, grains must be stored properly to prevent more mold toxins from growing. Of course, this rarely happens. Instead of being dried rapidly after harvest and sealed off from moisture, grains are often stored in less than ideal conditions, allowing moisture to accumulate and mold to grow. By the time the grain lands in your bowl, your breakfast cereal has a 42 percent chance of containing OTA.

Another issue with grains is the improper handling of equipment between crops. When one crop is harvested with the same equipment that was used on a moldy crop, they both become infested. This is such a big issue that animal ranchers track the global incidents of different mold toxins in grains because moldy grains make animals sick and reduce profits.

Ranchers have learned that feeding moldy grains or hay to animals makes them infertile and causes reproductive diseases, heart disease, and neurological diseases. So animal producers do what you'd expect Big Agriculture to do—they feed clean, mold-free grain to young or pregnant animals and feed the cheaper, moldy grain to animals in the last few months before they are slaughtered. This saves them money because the animals don't have a chance to get too sick before they're killed and we eat them.

That's why it's not enough for you to just avoid grains—though that's a good start. Try to avoid things that eat grains, too. Mold toxins accumulate in the milk of cows that are fed contaminated grains.[19] In fact, grain-fed animal products are even more likely to contain mold toxins than the grains themselves because grain-fed animals accumulate mold toxins in their fat. Conventional pork and poultry products are especially likely to be tainted with OTA. This is one reason I recommend eating grass-fed beef over poultry and only pork products from pastured, heritage pigs. You simply don't want a little dose of mitochondrial poison every time you eat.

COFFEE

Full disclosure: I write this section as the CEO of a coffee company that sells mold-free coffee beans. You can dismiss all of this information on those grounds (hehe), or realize that I believe so much in this research that I invested years of my life and most of my savings into creating a product that offers a safe alternative for everyone. It's your call, but I hope you'll keep reading regardless because this information is valuable and it can help you, even if you never try any of my products.

Mold toxins are a real problem when it comes to coffee.[20] A 2003 study showed that more than 90 percent of green coffee beans were contaminated with mold before processing,[21] while an earlier study revealed that almost 50 percent of brewed coffees are moldy.[22] These findings were so alarming that many countries rushed to set strict limits for the allowable amounts of OTA in coffee. *But the United States has no standards for coffee*, so it lags behind the EU, South Korea, Singapore, and China, all of which have set limits at economically feasible levels. And the coffee they reject actually gets sent to the United States for unsuspecting people to drink. In fact, a former president of the Specialty Coffee Association of America told me on video that he was present when a Japanese trade minister rejected one thousand containers of African coffee because of mold contamination. I asked him what they did with the coffee. He said, "They sent it to the U.S."

On average, decaf coffee contains even more mold toxins than caffeinated, partly because producers use lower-quality beans to make decaf. This is also because caffeine acts as a natural antifungal defense mechanism for the plant that deters mold and other organisms from growing on the beans. When you remove the caffeine, the beans are left defenseless against mold that forms if the beans are stored improperly after roasting.

Mold toxins form in coffee because the coffee industry saves money by harvesting coffee cherries and then letting them sit in unfiltered water for a couple of days to soften the pulp around the seed. During that time, uncontrolled fermentation creates mold toxins. The good news is that this method, dubbed "washed coffee," actually makes fewer mold toxins than the old "natural process" method, but it still makes more than you want to feed your mitochondria. You'd think that roasting the coffee beans would kill the mold, and it does . . . but it doesn't destroy the OTA chemical that the mold has created.[23]

What all of this means for you is that the type of coffee beans you purchase is incredibly important. Cheaper types of coffee cost less because they contain lower-quality beans that are more likely to be moldy. However, even the fancier organic coffees that you'd expect to be free of mold are still made with the "washed coffee" method that creates toxins. That's why Bulletproof coffee beans are made without fermentation, and that's why I test them for twenty-seven different toxins before deeming them safe for consumption.

I've lost track of the number of people who have told me something like, "I can't drink most coffee because it makes me jittery and anxious or tired, but I can drink yours without a problem!" I'm the same way. When OTA and other mold toxins in your coffee inhibit your energy production, you can expect to feel awful. On the other hand, when you get mitochondria-friendly polyphenols from your coffee without a load of mitochondria-damaging mold toxins, you get a mitochondrial upgrade!

If you're stuck without lab-tested coffee, here is how to reduce your risks of getting mold toxins in your coffee. First, look for single-estate coffee. That means the beans come from one place, so if you're lucky enough to get mold-free beans, you don't have to worry about them being mixed with other moldy beans. This is why blends of coffee are a bad idea, even if they taste good. Second, look for washed coffee, because washed coffee is better than natural-process coffee. Steer clear of natural process entirely. The third thing to do is to look for Central American coffee, which is often better than coffee from other regions. The fourth thing to do is to look for high elevation, as that can reduce mold problems by making stronger plants.

Remember, an "organic" label means nothing—most of the best coffees come from small plantations that could never afford an organic certification because the paperwork cost would put them out of business. Plus, organic coffee can sit in dirty water and grow mold toxins just like conventional coffee can.

If you follow these guidelines, you'll probably end up spending about $20 for a pound of high-end coffee. The bad news is that you still can't be sure it's low in mold. I created the Bulletproof process after throwing away far too many expensive bags of coffee because I couldn't drink them. But the good news is that if you get a clean bag of coffee, you won't go through it as quickly if you add butter and Brain Octane Oil as previously described. People naturally drink less coffee when it's free of milk and blended with good fats, so you'll end up saving money by drinking less coffee but spending more for the good stuff. It evens out in the end, and that's not even counting the savings of using Bulletproof Coffee to replace your breakfast!

DRIED FRUIT

Dried fruit contains even more sugar than regular fruit per ounce, and the drying process often creates high levels of mold toxins. Dried raisins, figs, dates, and plums generally have the highest levels of mold. If you get a headache, feel tired, or get sugar cravings after eating some dried fruit but not others, now you know one of the reasons. It's your mitochondria! Some dried fruit is sprayed with chemicals such as sulfites that harm your mitochondria, too. Depending on your unique biology, sulfites can inhibit lung and liver energy production and can deplete glutathione, especially if you already have toxic metals such as mercury in your system.

WINE AND BEER

So sorry to have to say it yet again, but these beverages are not brain boosters. Fifty percent of Mediterranean wines are contaminated with OTA.[24] The grapes become contaminated during the crushing process, and the toxins carry over into the wine. Beer, of course, gets its OTA from grains. While fermentation does decrease the concentration of OTA in the beer, it does not eliminate it. Wines produced in Europe are often lower in toxins than domestically produced wines, however, because government regulations abroad are stricter than those in the United States. I maintain a list of the best, cleanest wines I can find at bulletproof.com/wine. Beer, unfortunately, does not make the cut because of grain residue and gluten contamination.

CHOCOLATE

Like coffee, dark chocolate is a double-edged sword: it's a great source of polyphenols but can also be a significant source of mold toxins. One study found OTA on 98 percent of chocolate samples, with the highest amounts in bitter, powdered, and dark chocolate.[25] As you read in chapter 5, I prefer European chocolate because it undergoes more rigorous regulation, and I test the chocolate I use in my own products to standards in excess of European levels. If you feel odd after you eat chocolate, this could be your mitochondria telling you something. That's assuming you went for the 85% dark chocolate. If not, it's your mitochondria telling you to stop eating so much sugar in your chocolate!

NUTS

All nuts, especially peanuts (legumes), are likely sources of mold toxins. The nuts with the lowest risk are those you purchase still in their shells, but who has time to crack a bunch of nuts? I recommend buying whole nuts with the skin (not shell) still on, as manufacturers use damaged nuts that are far more likely to contain mold toxins to make slivered, chopped, or ground nuts, nut butters, nut flours, and even nut milks. Bonus points if the nuts are stored in the fridge at the store.

CORN

In a study that looked at 275 samples of corn, rice, and corn products, over a quarter of the corn samples were found to contain amounts of

OTA that were higher than the European limit.[26] This is especially problematic because corn is one of the most popular ingredients in the American diet today and is found in everything from artificial sweeteners to aspirin. Of course, it is also fed to nearly all factory-farmed animals, which then accumulate mold toxins from the corn they eat, and the toxins end up in their fat.

The most common corn fungus is called fusarium, which creates a toxin that inhibits mitochondrial function.[27] Because the industrial farming complex has been treating our soil heavily with antifungals for years, the fungus now lives on the roots of corn plants instead of on the corn kernels, making it impossible to see with the naked eye. But this mold literally injects its toxins into the roots of the plant, poisoning the whole thing—including the part we eat. On the Head Strong program, you will reduce the amount of corn in your diet to protect your mitochondria from these toxins.

Corn used to be a good gluten-free alternative. But when we started killing our soil with the herbicide glyphosate, we made corn soil fungi hyperaggressive, and now we pay for it by not being able to enjoy popcorn without the threat of toxins.

ARTIFICIAL SWEETENERS, FLAVORS, AND ADDITIVES

Manufacturers add all sorts of chemicals to processed foods in order to make them last longer and taste better. They are often found in products that are inflammatory to begin with, and they have their own harmful effects on the brain, making these foods twice as likely to cause inflammation, brain fog, and cognitive decline. You should avoid them at all costs because they universally lower your performance.

MONOSODIUM GLUTAMATE (MSG)

Of all of the toxins that manufacturers add to food, MSG might have the greatest impact on your cognitive performance. The glutamate in MSG is an excitatory neurotransmitter, meaning that it makes your neurons more likely to fire. When your neurons get too much glutamate from MSG, they continue firing rapidly for no good reason. This is called excitotoxicity, and it leads your neurons to run out of mitochondrial energy, create free radicals, and then die.

Many people suffer from headaches, brain fog, and even migraines after consuming MSG. Aren't those symptoms of mitochondrial problems? Yes! MSG damages mitochondria, and we've known that for years. In a 2003 study, researchers who exposed rats to MSG saw that it caused oxidative stress in vulnerable brain regions. This led to mitochondrial function impairment, which the researchers realized was an important mechanism of chronic neurodegeneration.[28] In other words, MSG makes you weak and dumb—the opposite of Head Strong.

In the United States, food ingredients that are less than 75 percent MSG do not have to be listed as MSG on the ingredient label. Manufacturers love to sneak things in like "spice extractives," "yeast extract," or "vegetable protein" because these contain MSG but you probably won't notice the name. If you're not sure about an ingredient, google an ingredient name together with MSG and you'll know right away if the manufacturer is trying to trick you.

ASPARTAME

The artificial sweetener aspartame is made up of two amino acids, or protein building blocks. One of them, phenylalanine, is chemically altered to form free methanol (wood alcohol). Free methanol is neurotoxic and is converted into formaldehyde in the liver.[29]

Formaldehyde is mitochondrial poison. A study from 2015 shows that it causes oxidative stress, greatly reduces cellular energy production, and eventually leads to apoptosis (cell death).[30]

Aspartame is also known as an excitatory neurotoxin because it causes your synapses to fire repeatedly. As you learned in chapter 3, your neurons are full of mitochondria because firing takes up so much energy. When your neurons fire relentlessly because you ate a man-made chemical, you're taxing your mitochondria at the exact same time that you're poisoning them.

In a recent interview with neurologist Dr. David Perlmutter, we talked about how aspartame and other sweeteners damage your gut bacteria, which causes brain inflammation. Chemical manufacturers aren't required to test what their concoctions do to bacteria in the gut, even though it can have a big impact on your brain.

SOY SAUCE

Your favorite sushi condiment is fermented with a type of fungus called aspergillus. Many species of aspergillus contain citrinin, a mold toxin that induces apoptosis.[31] Soy sauce also contains a stimulating neurotransmitter called tyramine, which causes oxidative stress, excitotoxicity, and damage to mitochondria.[32] Even worse, soy sauce is high in histamine, a stimulating neurotransmitter that can also cause systemic inflammation and mitochondrial slowing. Soy sauce also contains naturally occurring MSG, which in combination with tyramine can often cause migraines, brain fog, and food cravings. Soy sauce also contains gluten, and even the gluten-free version still has the tyramine and histamine problems.

The increase in histamine is a big problem for some people. One member of the anti-aging nonprofit I run heard me talk about this problem and went soy sauce–free. Within a week, her chronic hives and allergies went away!

I recommend you stop dunking your gorgeous salmon sashimi in soy sauce and try a sprinkle of sea salt instead! It's worth a try to see how you feel, and it's possible that you tolerate histamine and tyramine well enough to occasionally enjoy soy sauce with no food cravings later. But you may be surprised at how much better you feel without it.

NEUROTOXINS

These compounds are straight-up mitochondrial kryptonite that can destroy your performance even in very small amounts. Dr. David Bellinger, a professor of neurology at Harvard Medical School, estimates that Americans have collectively lost forty-one million IQ points just from ingesting neurotoxins.[33] Here are the most common sources of these brain poisons.

FLUORIDE

There is no scientific evidence to support the commonly held belief that adding fluoride to our drinking water or taking fluoride supplements is safe. Until the 1950s, doctors prescribed fluoride to reduce thyroid function. Yes, you read that correctly. Fluoride reduces the thyroid hormones that your mitochondria need to function and that your neurons need to maintain healthy myelin. Just 2 milligrams of fluoride a day is enough to reduce thyroid function. Now think about the fact that people who drink fluorinated water consume on average between 1.6 and 6.6 milligrams of fluoride a day. Unsurprisingly, a 2015 study showed that people living in an area with fluoridated water were 50 percent more likely to suffer from hypothyroidism (low thyroid function) than people in a nonfluoridated area.[34] Remember that low thyroid function changes the shape of your mitochondria and makes them less efficient!

Find out if your tap water is fluoridated, and then decide whether or not to keep drinking it. If your tap water is fluoridated, you can filter it to remove the fluoride, but not all filters successfully remove it. Look for a whole home filtration system that is effective at specifically removing fluoride.

The link between cavities and fluoride is tenuous at best. But even if fluoride were the world's best cavity prevention tonic (it's not), I would gladly risk a few cavities in order to have rock star mitochondria that reliably gave me more energy than I knew what to do with!

GENETICALLY MODIFIED ORGANISMS (GMOS)

There are plenty of environmental reasons to avoid consuming genetically modified foods, but for our purposes, I'll focus on their neurotoxic effects. GMOs are almost universally sprayed with Roundup, which is part of a class of pesticides known as organophosphates. The United States Environmental Protection Agency (EPA) lists organophosphates as *acutely toxic* to bees, wildlife, and humans. Even low levels of exposure have been linked to adverse effects in the neurobehavioral development of fetuses and children. When babies are exposed to low levels of these pesticides in the womb, they have lower IQs and experience lifelong problems with learning and memory.[35]

So why the heck is this being sprayed on our food as it grows and right before it's harvested?

Organophosphates irreversibly inactivate an enzyme in your body that breaks down the neurotransmitter acetylcholine. As you read in chapter 5, you need *some* acetylcholine to stimulate your muscles and schedule REM sleep. The problem is that organophosphates destroy your body's ability to maintain proper amounts of this neurotransmitter. When your body can't get rid of excess acetylcholine, you get muscle tension and synaptic overload.

Organophosphates are also terribly destructive to your mitochondria. They alter all five mitochondrial complexes, disrupt the mitochondrial membrane, reduce ATP production, mess with antioxidant cellular defenses, and promote cell death.[36]

How can you avoid GMO produce? Your best bet is to buy organic at the grocery store or get to know a local farmer and buy directly from him or her. You can also go online and learn more about which fruits and vegetables are never GMO—like one of my favorites, avocado. Does it take a little more work (and money) to seek out non-GMO produce? Sometimes it does. But keep in mind that every time you eat a GMO food, you're being exposed to a small dose of organophosphate. I think it's worth the time and investment not to expose your mitochondria to poison every day!

MERCURY

Mercury is a heavy metal and one of the most toxic of all in its class. It depletes the antioxidants that your mitochondria need to combat oxidative stress,[37] leading to inflammation, cell damage, and mitochondrial dysfunction. It has also been directly linked to reduced IQ.[38] Thanks to water pollution, mercury is commonly found in our seafood. Mercury accumulates in the tissues of fish, so the higher a fish is on the food chain, the more likely it is to contain dangerous levels of mercury. The highest concentrations of mercury are found in tilefish, swordfish, shark, and mackerel. You will avoid these fish on the Head Strong program. There is some evidence that seafood contains selenium, which can help to counteract the mercury in seafood, but it's still a better bet to avoid the most mercury-tainted species.

SUGAR

Yes, I consider sugar at commonly consumed levels to be a neurotoxin. It is part of nearly every major degenerative illness, including (and especially) Alzheimer's disease—so much so that many doctors have begun referring to Alzheimer's disease as "type 3 diabetes."

Scientists have known that sugar is bad for the brain since 1927, when a biochemist named Herbert Crabtree discovered that elevated glucose levels lower mitochondrial function. This is called the Crabtree effect. To this day, we are constantly learning more about exactly what sugar does to our brains. A 2013 article in the *New England Journal of Medicine* stated that in diabetic patients, even mild elevations in blood sugar were strongly related to the development of Alzheimer's disease.[39]

Insulin resistance is also linked to Alzheimer's. Insulin resistance—sometimes referred to as prediabetes—is a condition in which the body becomes less sensitive to the presence of the hormone insulin. As you might know, the pancreas produces insulin to help metabolize sugar. When too much sugar is present in the blood (the result of consuming a lot of sugar), too much insulin is produced in response. Over time, this onslaught of insulin desensitizes the body to smaller amounts of insulin, so your pancreas will continue to churn out insulin in huge amounts even when you eat small quantities of sugar.

What does this have to do with your brain? Insulin helps facilitate communication between your neurons. When you're insulin resistant, the excess insulin rushes to your brain, and important messages get lost in the flood. A 2015 study showed that people who were insulin resistant (but who did not have Alzheimer's disease or diabetes) scored lower on memory tests than those who were not insulin resistant. In another study conducted that same year at UCLA, rats that were fed a high-sugar diet for six weeks experienced declines in their ability to navigate through a maze and exhibited less synaptic activity than rats that weren't fed sugar. Their neurons literally couldn't signal to each other, and the rats lost their ability to think clearly or complete tasks they'd learned just six weeks earlier.[40]

In humans, sugar has been shown to make us moody and angry by messing with our neurotransmitters and decreasing the number of dopamine receptors in our brains. This makes it harder to feel the effects of dopamine and creates dopamine resistance, which is the same neurological response that is observed in drug addicts. Sugar is as powerful as a drug when it comes to its effects on our brains!

But, as you read in chapter 4, perhaps the greatest havoc sugar wreaks on the brain is in the form of inflammation. When you have high levels of blood sugar and insulin, your body releases inflammatory cytokines. This can create a vicious cycle, as insulin causes inflammation and inflammation causes more severe insulin resistance. Blood sugar levels then creep higher and higher as you become increasingly inflamed, foggy, forgetful, and tired.

All forms of sugar are bad for your brain, but fructose—found in fruit, high-fructose corn syrup, and agave nectar—is the worst. Fructose creates oxidative stress[41] and feeds the bad bacteria in your gut, leading to even more inflammation. Fructose is implicated in damaging mitochondria in skeletal muscle cells, harming the mitochondrial membrane, and impairing cellular respiration and energy metabolism.[42] While your brain won't suffer too much if you eat moderate amounts of *whole*, seasonal fruits, you should avoid consuming excessive amounts of fructose and completely stay away from fruit juice and foods that contain high-fructose corn syrup and agave nectar. I recommend no more than about 20 grams of fructose a day from any source for maximum cognitive function.

ALCOHOL

In addition to all of the other ways that alcohol is bad for your brain, it causes oxidative stress in your mitochondria while simultaneously reducing your mitochondria's oxidative stress defenses. This is like making the good guys weaker so they can't fight off the bad guys while adding more bad guys at the same time! It leads to a vicious cycle of more and more cell damage.[43]

This added oxidative stress also makes your cells more susceptible to apoptosis, or death. Basically, alcohol slows down the energy production in your cells, weakens them, and then makes them more likely to die. Sure, having a beer or two with your friends is fun, but there are other ways to have fun without literally killing your brain cells.

Keep in mind that many flavored alcohols (and mixers) contain high-fructose corn syrup, which is bad for your mitochondria even without the alcohol. And beer and wine are unfiltered and un-distilled, so all of the fermentation by-products are still present. This includes mold toxins such as OTA.

On the Head Strong program, I recommend that you abstain from alcohol completely for two weeks. After that, choose distilled clear spirits (not beer) or, if you tolerate it, low-toxin wine from the vendors listed at bulletproofexec.com/wine. You can take some vitamin C and glutathione supplements when you have alcohol to help your mitochondria survive a few drinks.

Is Coconut Oil All It's Cracked Up to Be?

Like butter, coconut oil once had a bad reputation due to its extraordinarily high concentration of saturated fat. Fortunately, we now know that there's no association between coconut oil and heart disease. I like coconut oil because it is a widely available, good saturated fat that is relatively stable at high temperatures (which means it's great for cooking), and it tastes good (in some recipes). More recently, coconut oil has earned a reputation for being a brain-boosting fat source. So which is it—junk food or super food? The truth is somewhere in the middle.

Coconut oil naturally contains eleven different types of fat, each of which does something different in your body. Only four of those types of fats are categorized as medium-chain triglycerides, each with very different metabolic effects. The most common (and cheapest) of these is lauric acid—a fat that's technically classified as an MCT because a chemist decided that long before we knew how the body used it. The problem is that your body processes lauric acid as if it were a long-chain fat (remember: the shorter the fat, the better). Thus it does not raise ketones like other true biological MCTs, and there is a strong case that lauric acid should no longer be classified as an MCT. Coconut oil is about 50 percent lauric acid. That's a little scary, because a recent study found that lauric acid can cause immune T-cells to create more inflammation and (in mice) make the neurodegenerative disease MS worse.[44] This is not a call to stop eating coconut oil, but it means you should use it in moderation (two to three tablespoons per day) and eat it with vegetables. You should never add extra lauric acid to your diet on purpose, even when it's marketed as an MCT oil.

The remaining three types of MCTs in coconut oil are caproic acid, caprylic acid, and capric acid. Caproic acid is present in small amounts in coconut oil. It tastes bad, and it often causes stomach upset, but it does raise ketones. If you use generic MCT oil that is not triple distilled, you may get traces of this oil and feel it as a burning in your throat (and disaster pants later). Caprylic acid is the rarest MCT in coconut oil—4–6 percent of its fat. It has potent antimicrobial properties to help you have a healthy gut and provides the most ketones for your brain of any other oil. Capric acid makes up about 9 percent of the fat in coconut oil. It takes slightly longer for it to turn into energy for your brain, and your ketones do not rise as high, but it is more affordable and widely available than caprylic acid.

Each of the MCTs in coconut oil can be distilled to create separate oils, either bottled individually or as blends. Some of these generic MCT oils do raise ketones more quickly than coconut oil itself. The problem with many of these oils, however, is that low-cost distillation methods may allow traces of caproic acid to remain, so you get disaster pants. Another big issue with generic MCT oils is that, like coconut oil, they can contain a high percentage of lauric acid, which is associated with inflammation. This dilutes the power of the other MCTs in the oil and makes it less effective at raising ketones. The least ethical companies actually try to sell this as a benefit to unwitting consumers.

I created two oils to solve this problem: Brain Octane Oil, introduced in chapter 5, and XCT oil. Both oils are triple distilled in the United States on food-grade machinery, using only coconut oil as the source, with no solvents ever. Generic MCT oil is often made overseas with single distillation of solvent-extracted palm oil, which results in more oil impurities, and palm oil usage kills orangutans and destroys rain forests. XCT Oil is distilled to be about six times stronger than coconut oil and is a more affordable mix of capric and caprylic acids. It does not raise ketones as much as Brain Octane Oil, and there is a limit to how much you can have before it causes gastric distress. But XCT Oil supports the gut biome and helps lower inflammation, which can reduce cravings and brain fog. Brain Octane Oil, made of carefully filtered caprylic acid, is the single most powerful form of MCT and raises your ketones the highest and the fastest, far beyond any other fat found in coconut oil or any other food. Caprylic acid comprises 5–6 percent of the fat found in coconut oil, so you'd have to eat more than a dozen tablespoons of coconut oil to get the amount of caprylic acid in just 1 tablespoon of Brain Octane Oil. It requires the most coconut oil to manufacture, but it is far more potent than XCT oil, generic MCT oil, or coconut oil, and most people can have a lot more of it without the digestive problems of generic MCT oil.

A recent groundbreaking study looked at the effects of intermittent fasting with coconut oil versus the effects of intermittent fasting with two subtypes of MCTs, the ones found in XCT oil. When subjects fasted and then consumed coconut oil, their ketone levels remained static—the coconut oil did not raise their ketone levels at all compared to fasting. When they fasted and then consumed a combination of coconut oil and other MCTs, their ketone levels went up slightly. But they got by far the biggest boost in ketones after fasting and then consuming Brain Octane Oil.

The most important thing about Brain Octane Oil is that it will raise your ketones even if you are eating carbohydrates. Until now, you would have had to go on a low-carbohydrate diet for four days to get the type of ketone energy you get when you add Brain Octane Oil to whatever food is on your plate (or in your coffee!).

So don't fall for the marketing hype and eat coconut oil because it's a good source of MCTs—it's not. But it is a great source of saturated fats. A lot of people will skimp on their Bulletproof Coffee by adding coconut oil or generic MCTs. If you do this, you won't be harming yourself—and I won't judge you for it. But you need to know that you're not getting the same ketone or brain benefits that you would with Brain Octane Oil. And you might get disaster pants from the MCT! The choice is yours.

GOOD FATS GONE BAD

Even if you carefully avoid all of the foods you've read about in this chapter so far, it's very likely that you're transforming your healthy, well-chosen foods into ones that slow your mitochondria with one common mistake: your cooking method.

When you smoke, fry, or grill meat, you create two carcinogens: heterocyclic amines (HCAs) and polycyclic aromatic hydrocarbons (PAHs). These compounds do more than cause cancer, though. HCAs are neurotoxic and induce tremors. In one study, patients with essential tremor, a common neurodegenerative disease, had 50 percent more HCAs in their bloodstreams than people without tremors.[45]

Even worse, both HCAs and PAHs are known to inhibit mitochondrial function. When monkeys were injected with HCAs, they experienced mitochondrial degeneration.[46] When HCAs were given to rats, their mitochondria became mutated and enlarged.[47] Unfortunately, as you now know, bigger isn't better when it comes to your mitochondria.

Similarly, treatment with PAHs causes oxidative stress, mitochondrial function defects, and mitochondrial damage.[48] But then it gets worse. When human bronchial cells were treated with PAHs for only four hours, the PAHs prevented the cells from dying during mitochondria-induced apoptosis.[49]

Apoptosis is a funny thing. You don't want your healthy cells to die, but you definitely *do* want your unhealthy ones to die. When cells are damaged from oxidative stress and then refuse to die during apoptosis, these damaged cells often replicate and cause even more damage. Many scientists now believe this is one of the major mechanisms behind cancer growth and one reason PAHs are so carcinogenic.

Another problem with some cooking methods (which are outlined in detail below) is that they damage important proteins. When a protein becomes damaged it is considered "heat denatured." Denatured proteins aren't necessarily toxic (you also denature proteins as you digest your food), but they also can't do their jobs very well. Whey protein, for example, boosts your mitochondria's production of glutathione, an incredibly important antioxidant, but it cannot perform this important task as well when it is denatured.[50] A lot of recipes for grain-free baked goods include whey protein, but I avoid them for this reason.

But the worst thing you can do when it comes to cooking your food is to damage the fats they contain by oxidizing them. Fats, especially unsaturated fats, are very sensitive. As you read earlier, they are easily damaged by exposure to heat and even light. Below I discuss specific cooking methods that cause these problems, but first let's take a look at what these damaged fats do to your body.

When you consume damaged fats, your body still uses them to create cell membranes. As you know, your brain is made mostly of fat. Your myelin is made mostly of fat. Your hormones are made mostly of fat. And your mitochondria rely on fat to function. When the cell membranes in your brain and the rest of your body are made of damaged fats, they are less flexible and less functional. Your neurons can't send or receive messages as efficiently, and your mitochondria start to degrade. Oxidized fats also disrupt hormone and neurotransmitter signaling. One way they do this is by producing excessive amounts of glutamate, the same excitatory neurotransmitter in MSG that can cause neurons to die from excitotoxicity.

But the biggest problem with oxidized fats is that they are highly inflammatory. Every time a damaged fat molecule is used as a building block in the body, it creates oxidative stress. As you read in chapter 5, polyunsaturated fats are the type of fats that are most easily damaged. When heated, these oils produce compounds called dicarbonyls that are particularly toxic to the brain. They damage mitochondria and cause oxidative stress.[51] Dicarbonyls are also precursors to advanced glycation end products (AGE), which, as you read earlier, cause inflammation and further compound oxidative stress.[52]

This is probably one of the main reasons I started to feel so lousy after losing fifty pounds on a low-carb, high-fat, ketogenic diet. I was eating foods like pork rinds and consuming toxic chemicals like aspartame all day long, and while I was losing weight and staying in ketosis, I was also unwittingly creating a lot of inflammation in my body. I've seen this over and over with ketosis diets that allow inflammatory foods and chemicals. The *type* of fat you eat matters!

Luckily for me, we replace half of the fat in our cells every two years. Those damaged fats are now long gone from my cell membranes, and I can feel the difference in my brain. I still eat a lot of fat, but I don't cook with it at high temperatures. When I do cook with fat, I use lower temperatures, water or steam, and a lot of antioxidant spices to counteract the oxidative stress. This is something that not many people think about but that can make a huge difference in your daily performance.

Here are some of the top offenders when it comes to cooking fats.

FRYING

Whether it's French fries, fried chicken, fried fish, or fried Snickers bars, anything fried is full of damaged fats. The process of deep-frying bathes your food in oxidized fats and denatured proteins. The high temperature used during deep-frying compounds the toxicity of your food by producing PAHs and HCAs. You already know that fried food is bad for your waistline, but now you have another reason to avoid it: it's bad for your brain! It's worse if you're eating restaurant fried food, because they use the same oil for longer periods, and it gets increasingly damaged as time goes on.

SAFFLOWER AND SUNFLOWER OILS

Safflower oil is heated to high temperatures to isolate the oil, but this oxidizes the fragile compounds in the oils. Sunflower oil has the same problem as safflower oil, but it is even more prone to oxidation and it has a lower smoke point. This pretty much guarantees that the sunflower oil you eat is oxidized even before you cook with it. Avoid both of these oils.

VEGETABLE OILS, SOY OIL, CORN OIL, AND TRANS FATS

You already know to stay away from these types of fats. The fact that they're so easily oxidized is just one more reason. Soybean oil turns on your genes for inflammation and interferes with mitochondrial function.[53] In rats, corn oil suppresses mitochondria and is linked to colon cancer.[54] Use grass-fed butter instead!

BARBEQUED MEAT

You didn't feel foggy and hungover after your Fourth of July BBQ just because of the margaritas you drank. When the fats in your delicious meat hit the open flame, they were converted into carcinogenic and inflammatory HCAs and PAHs. Most barbecue sauces are also full of sugar. You'll perform better all summer long if you forgo eating charred meat. If you're going to grill, wrap your meat in foil first!

Vegetarian Omega-3s—Healthy Fat or Not?

I have nothing against vegans or vegetarians. I was a raw vegan myself for a period of time. But after being on that diet for a while, my brain function really began to suffer, and my lack of sufficient EPA and DHA (the two types of omega-3 fatty acids that your brain needs most) was probably one reason why.

EPA and DHA are found exclusively in seafood and marine algae. Many vegetarians try to meet their omega-3 needs by supplementing with alpha-linolenic acid (ALA), which is a precursor to both EPA and DHA. This means that your body uses ALA to make omega-3s. ALA is found in many seeds including flax, hemp, and pumpkin seeds. This is why many vegetarians supplement with flax oil.

The problem with these supplements is that your body is not very good at using ALA to make EPA or DHA. In fact, you convert less than 5 percent of the ALA you consume into EPA, and you convert even less (a mere 0.5 percent) into DHA. To make matters worse, your body uses iron to convert ALA into these small amounts of EPA and DHA, and many vegetarians and vegans are already low in iron. This depletes their iron stores even more. These supplements probably won't hurt you, but the conversion rate is so low and uses up so much of your body's precious energy and iron reserves that it really isn't worth it.

The information in this chapter is not meant to scare you. Rather, I want you to understand the huge impact that food can have on your brain. This gives you a lot of power because you are in charge of everything that you put in your mouth! That means that you get to decide how many inflammatory, toxic foods you want to feed your mitochondria and how good (or bad) you want to feel.

You don't have to make perfect choices at every single meal, every single day. Sometimes it might be worth it to eat something that you know will leave you a little bit inflamed. You can decide when you're willing to take the hit. But if you have a big interview or presentation the next day, it's incredibly empowering to know that you can control your mental performance just by making a smarter choice at the dinner table. When you start choosing anti-inflammatory foods that are unlikely to be contaminated with chemical or naturally occurring toxins, you will feel an immediate difference in your mental performance and clarity. This is one of the most important steps to becoming Head Strong.

Head Points: Don't Forget These Three Things

- Dairy protein, gluten, trans fats, and vegetable oils cause inflammation in everyone.
- Mold toxins are particularly toxic to your mitochondria and are commonly found in grains, coffee, dried fruit, wine, beer, chocolate, nuts, and corn.
- You can damage healthy fats by cooking them at high temperatures. This makes them toxic.

Head Start: Do These Three Things Right Now

- Stop using artificial sweeteners. They are toxic to your mitochondria even in small amounts. While you're at it, cut down on your sugar intake, particularly the fructose that is found in fruit juice and high-fructose corn syrup.
- Never eat fried food! Frying damages fats and makes them toxic.
- Buy organic produce whenever possible, as GMOs are commonly sprayed with toxic pesticides.

7

AVOID TOXINS AND IMPROVE YOUR BODY'S DETOX SYSTEMS

The term "detox" has become ubiquitous in health and wellness circles in recent years, and it's often invoked to describe an array of (sometimes bizarre) diet regimens that promise big results. But what *is* detoxing? And just what kinds of toxins are hanging around in our bodies?

Detoxing can take many forms, but when done correctly, it is an essential part of any routine. Because of the impact that toxins can have on your brain, detoxing is especially important when it comes to optimizing brain performance. The truth is, the body has its own innate detox system and is able to remove some toxins on its own—but it would be naive to assume that our bodies can process and completely remove all of the chemicals we are exposed to daily.

From naturally occurring environmental toxins like mold to man-made toxins like lead-based paint, we are assaulted by a greater number of hazardous substances than our bodies can reasonably handle. Unfortunately, removing these toxins from your life isn't as easy as simply avoiding the foods you read about in chapter 6. Your body is assaulted with a daily dose of toxins from nonfood sources as well—they can be found in the air you breathe, the nooks and crannies of your home, and even your medicine cabinet. You likely aren't even aware of their presence, but your brain certainly is because these toxins stress your mitochondria.

Of course, in every area of life, there is good stress and bad stress. Sometimes the saying is true—what doesn't kill you makes you stronger. We learn new things during times of mental stress. Our muscles grow when we exert the right kinds of physical stress. And when it comes to the brain, the right kinds of stress can help to strengthen mitochondria, fuel mitochondrial growth, and kill off damaged mitochondria that are no longer useful. But when toxins stress your mitochondria the wrong way, nothing good happens. Toxic stress does not lead to growth or renewal—just the opposite. It kills or damages healthy mitochondria.

As you know, your neurons are dependent on your mitochondria for energy, so when you have dysfunctional mitochondria, you have dysfunctional neurons. It's a vicious cycle that you can end by avoiding toxins and helping your body to detox from the ones that are already hurting your brain.

ENVIRONMENTAL MOLD

Earlier in this book, you learned that mitochondria are microscopic bacteria that power our cells and influence how we respond to our environment. And as we saw in the previous chapter, bacteria have been locked in a mortal struggle with mold since the dawn of time, each producing toxins to inhibit the other. So is it any wonder that mold, and the toxins it produces, could hinder the function of your mitochondria?[1]

Mold creates toxic chemicals, or mycotoxins, at levels that are typically far too low for you to smell or taste—but that doesn't mean they aren't making their way into your body through your lungs and skin, not to mention your food and water. Even scarier, some species of mold make different toxins that act synergistically, meaning each type of toxin amplifies the effects of the other. According to a major research paper about mold and mitochondria, "there are no safe levels of mycotoxins."[2] I agree with this, but I also acknowledge that it's impossible to completely avoid mycotoxins unless you live in a bubble. So I work to minimize my exposure when it's convenient, and accept low exposure when it's worth it. Paranoia is just another form of stress, and that's bad for your mitochondria, too!

The impact of mycotoxins on your health depends on a few factors:

the type and amount of mycotoxin to which you've been exposed, the duration of exposure, the other toxins present in your environment, and your personal health profile (age, sex, genetic background, etc.). Your diet also plays a key role in your susceptibility to these toxins. If you're missing certain vitamins, not eating enough, or abusing alcohol,[3] your mitochondria are already weakened and therefore more vulnerable to attack. If they then take a hit from mycotoxins, you're going to feel pretty awful.

About 28 percent of the population has a genetic sensitivity to mold. For these people, exposure to mycotoxins can cause a variety of symptoms that include brain fog, cognitive issues, fatigue, joint pain, nausea, weight gain, chronic sinusitis, and asthma. I am one of these people. As you read earlier, I grew up in a house that had a water-damaged basement, with mold growing behind our innocent-looking wood paneling (hey, it was the '70s). As a kid, I had strep throat every month. I had nosebleeds ten times a day, rashes, asthma, bruising, obesity, and arthritis at age fourteen. When I was tested for allergies, the results always came back negative. Finally, when I was sixteen, my tonsils were removed in an attempt to prevent me from getting another throat infection. I got my first sinus infection the following week.

I remember going to the doctor and saying, "I feel like I've been poisoned." No one thought to look in my environment for the source of that poison before prescribing antibiotics, which are actually derived from mold toxins. No wonder I felt like crap! I needed something to support my mitochondria, not to add more toxins to the assault.

Even if you fall into the 75 percent of people who are less sensitive to mycotoxins, you're not off the hook. Not only do mycotoxins cripple your mitochondrial function, they also trigger an inflammatory immune response. The more frequently you are exposed to mold, the more chronically inflamed you are likely to become. Eventually, your immune system becomes hypersensitive to the threat and responds to the presence of even a tiny amount of mold toxins right away. We call this response an allergy.

In addition to triggering an inflammatory response throughout your body, toxic mold exposure also causes the space between your cells to become wider. This leads your cell membranes and your blood-brain barrier to become more permeable. You do not want either of those barriers to be anything less than airtight. When cellular barriers become leaky, fluid, plasma, and other foreign particles can travel unrestricted to parts of the body where they don't belong—namely, your brain.[4]

Much like the way the blood-brain barrier protects the brain, a similar type of barrier controlled by a series of tightly locked cellular junctions protects the lining of your gut. As we learned in the previous chapter, these junctions are loosened when you are exposed to toxic mold, allowing foreign particles (like partially digested food) to leak out of the gut and enter the bloodstream. Of course, your body detects these foreign intruders as the hazards they are and sounds the inflammation alarm to attack them.

Over time, this inflammatory response becomes automatic. Your body feels like it's constantly under attack and mounts a counteroffensive against substances and foods that may have never bothered you before. Maybe you were previously able to eat gluten or dairy products without a problem, but now you feel terrible if you have even a tiny serving of either. That's because the permeability of the gut lining allowed particles of these foods to leak out, and after your body constantly launched an attack against these foods, you developed an allergy to them.

The 25 percent of people like me who are extra sensitive to mold toxins are the canaries in the coal mine for everyone else. Exposure to mold may affect us more dramatically, but if we're suffering, chances are you're suffering, too. Let me give you an example. Not long ago, I attended a professional conference in San Diego. One night, we were scheduled to go on a dinner cruise around the city. From the moment I stepped onto the boat, I noticed that it smelled like a mop—a dead giveaway that there was mold in the air. I considered heading back to the dock, but I really wanted to spend some time with the other conference attendees, so I decided to take the mitochondrial hit. Remember, once you know about the effect of toxins on your brain, exposure is largely within your control. In this case, I thought the upside would be worth the risk.

The next day at the conference, I struggled over and over again to come up with the right words as I was giving my keynote presentation. This problem used to plague me daily, and it made me feel stupid. But after years of hacking my mitochondria and upgrading my brain, it *never* happens to me anymore. I can rely on my brain to perform at a consistently high level. So my sudden inability to recall words was noticeable and embarrassing to me. (By the way, it's not "normal" for *anyone* to experience word recall problems—it's a symptom of impaired brain function.)

The day after that, I became really fatigued. I had stomach cramps, sugar cravings, and I was constipated. Another day later, my face broke out, I developed canker sores in my mouth, and I had a nosebleed for the first time in years. That night, I slept for twelve hours (compared to my usual five or six), and I still had no energy the next day. I felt like total garbage. I had also grown love handles from the inflammation. I had a chance to hang out with Marshall Goldsmith that day, a successful business author I've long admired, and we had our photo taken together. When I posted it online, one of my followers commented on the sudden appearance of my inflammation-driven man-boobs. Delightful!

While my body's response to mold may be more extreme than yours, most people experience some version of these symptoms without recognizing the source. Toxic mold exposure hits your brain mitochondria first, then your gut, and then your skin, and it can really damage your performance in the process. I wasn't the only one at the conference who felt exhausted the day after the cruise. Some people said their allergies were acting up and others blamed their fatigue and sluggishness on being hungover (even though they drank very little).

Consider this story to be a cautionary tale: if you're not performing well or don't feel like yourself, there's always a reason! When I interviewed brain disorder specialist and bestselling author Daniel Amen, MD, about the effects of toxic mold exposure, he told me that it's never normal to have impaired cognitive function. If you suddenly can't remember things, you shouldn't just dismiss it as a symptom of "getting older." Instead, pay attention to your body's signals, no matter how subtle—they could be signs of trouble ahead.

Dr. Amen told me that it's possible for mold exposure to impair your brain enough that if you were to take an IQ test before and after exposure, you could score 15 points lower after exposure. How crazy is that? Think about the "off days" you've had when you just couldn't focus and blamed it on not trying hard enough. What if it wasn't a lack of effort that impaired your performance but instead it was exposure to toxins that impaired your mitochondrial function?

Through the use of his SPECT brain scan technology, Dr. Amen is able to observe how mold affects the brain on a physical level. He explained that in the scans he's seen, mold exposure visibly damages the amygdala, the part of your brain that is involved in impulsive, reactive emotions such as fear, anger, and anxiety. An impaired amygdala can cause you to fly into a rage for seemingly no reason. This can devastate your performance, not to mention your relationships. Dr. Amen says that people whose brain scans reveal toxic mold exposure often hate themselves because they don't know why they're having such a difficult time controlling their emotions—that is, until they see the image of their brain looking visibly impaired.

Maybe you're thinking, *That's too bad, but it could never happen to me without me knowing it!* Well, here are some numbers that might surprise you: Close to 50 percent of the buildings in the United States contain water damage. That's largely because in the 1970s, we started making buildings out of drywall, which absorbs moisture. This creates an ideal environment for mold to grow. Around the same time, we tried to keep mold at bay by adding fungicides to our paint. Unfortunately, the mold outsmarted us, mutating and creating a fungicide-resistant species that makes even more mycotoxins.

It's also important to note that once mold grows in your home (or office or school), it quickly spreads to contaminate everything else around you—your clothes, your furniture, and all of your belongings. Carpets in particular are porous and will sop up toxic mold. This means that even if you pack up and leave a moldy house, you're literally just taking your problem with you unless you get rid of your belongings, too.

Despite all of the evidence to support how damaging toxic mold is—especially for the brain—most doctors don't learn about mold in medical school and aren't able to recognize toxic mold exposure in their patients. As a result, many people who are suffering from the symptoms of mycotoxin exposure are often dismissed, misdiagnosed, or treated for psychological illness instead of their very real physical illness.

I was fortunate to interview Dr. Scott McMahon, a leading expert in chronic inflammatory response syndrome (CIRS), which often stems from exposure to mycotoxins. Dr. McMahon estimates that at least half of his patients were told by their previous doctors that they were either crazy or imagining their symptoms. A huge number of them were prescribed Zoloft, while others were simply sent on their way. Understandably, many of them grew depressed and even became suicidal.

This is why it's so important for you to listen to your body and to know how to protect yourself—and your brain—from exposure to mycotoxins. Granted, sometimes this is easier said than done. But the following steps will help you avoid mold and recover from its toxic effects if you have been exposed.

- **Avoid Water-Damaged Buildings.** Waterlogged building materials are the perfect breeding ground for mold.

  Avoid living, working, or attending school in any building that has been damaged by water from a flood, broken pipes, condensation, or water leaks of any kind until a licensed mold specialist has remediated the building.

  Water stains on the walls or ceilings or a funky smell should be enough to tell you not to spend time in a building. If you enter a hotel room and notice a musty smell, ask for a new room. Schools, government buildings, and older structures are at high risk because they usually aren't well maintained. The number of school buildings that are interfering with our children's ability to learn thanks to mold is horrifying.

- **Prevent, Identify, and Repair Water Leaks.** Keep your drains, toilets, and pipes in good working order to prevent flooding. Have the pipes in your home professionally inspected for leaks. Leaks are often hidden behind walls or cabinets, so you can't tell they're there just by looking. Inspectors can use an infrared camera to find hidden leaks.

If you find water leaks, repair them immediately and dry the area quickly. When I had a recent leak in my home office, I dried the area right away and misted it with Homebiotic, a form of probiotic bacteria that compete with toxic mold. Creating balance as a preventive measure is a better strategy than waiting for damage to occur and then treating it with chemicals. Check baseboards, ceilings, and walls for soft spots, stains, or other signs of water damage. Also be sure to keep your roof well maintained. It will not only keep you dry, but it will also prevent you from having to deal with a very expensive mold remediation later.

- **Consider Safer Building Conditions and Materials.** When buying or renting a new home, pay attention to the building materials and the condition of the home and have it inspected for an intact moisture barrier in the walls. You may assume that newer homes are safer, but the opposite is actually true since the materials that have been used in construction since the 1970s are especially susceptible to toxic mold growth. Try to spend at least half a day in the space you're going to buy or rent. See how you sleep that night and how you feel when you wake up. If it's worse than normal, or if you notice any symptoms of cognitive decline, consider it a red flag. Make sure to have the home thoroughly tested for mold before you move in.

- **Ensure Proper Ventilation.** Buildings that have been vacant for a long period of time often contain humid, stagnant air. Mold grows in moist environments that lack proper airflow—rooms that you would probably call "stuffy." Make sure there is good ventilation in your home, office, and school. Poorly installed AC (air-conditioning) units and HVAC (heating, ventilating, and air-conditioning) systems in hot, humid climates also pose a risk. Inspect these carefully. Cool AC ducts cause condensation when they come in contact with moist air, and mold grows on the condensation.

- **Have Your Home Professionally Inspected.** If you are about to lease, rent, or buy a home, find a mold testing professional in your area to perform an ERMI (environmental relative moldiness index) air test. This test has been developed as a tool to evaluate the potential risk of indoor mold growth and its associated health effects. Mold testing companies can also perform a complete test of the interior and exterior of your home.

  Even if you've lived in your home for a long time, if you are experiencing unexplained symptoms like the ones you've read about in this chapter, it's a good idea to have your home inspected for mold.

- **Avoid Mold Through Removal and Remediation.** If testing reveals that you have toxic mold in your environment, remove yourself from the environment and work with a remediation specialist to develop a cleanup plan. Improperly removing toxic mold can cause the spores to become airborne. This can be even more dangerous to your mitochondria than asbestos or lead paint because the impact is faster and more severe. I've heard from countless people who had a few manageable symptoms of mold exposure but then grew very ill when contractors started mold remediation work on their house.

  No matter how strong your mitochondria are, there is no better strategy to avoid the toxic effects of mold than to avoid mold altogether! If you experience the symptoms of mold illness, contact a health care practitioner. Even if only one person in your home has symptoms, it is important that everyone takes precautions. Mold is toxic to everyone—it is just that some of us feel its effects more strongly than others.

- **Detox Your Diet.** The Head Strong plan will steer you away from foods that are most likely to contain mold and toward the ones that will help your body detox. Eating a diet low in sugar, rich in antioxidants and polyphenols, and containing high-quality fats, grass-fed protein, and organic, responsibly sourced foods is the best way not only to avoid foodborne molds but also to detox from mold exposure. Even if you're not in need of a mold detox, eating this way will supercharge your mitochondria and help you feel amazing.

- **Detox Through Supplements.** The mitochondrial-enhancing supplements in the Head Strong program will help you make energy, but you still need to get mold toxins out of your system as quickly as possible. You can do that by taking supplements that contain activated charcoal and bentonite clay, both of which aid in detoxification.

- **Detox from Candida.** Ingesting mold can stimulate an overgrowth of yeast in the body. Some yeast is always present in our bodies and is perfectly healthy. But when a type of yeast called candida takes root, it can wreak havoc, creating autoimmune, digestive, and cognitive issues.

  Antifungal medications and herbs can work wonders for candida. Grape seed extract, oregano, berberine, coconut oil, and regular use of Brain Octane Oil will all help to control yeast overgrowth. A good probiotic will help, too.

HEAVY METALS

Whether we're talking about music or toxins, heavy metal should have stayed in the 1980s.

Your mitochondria are very sensitive to heavy metals like lead, mercury, nickel, uranium, arsenic, and cadmium. Even a small amount of exposure to these toxins for a short period of time is sufficient to impair mitochondrial energy production and increase mitochondrial death.[5] In one study, only three hours of exposure to heavy metals caused significant mitochondrial dysfunction, and forty-eight hours of exposure caused a 50 percent decrease in energy production.[6] Heavy metal exposure also made the mitochondrial membrane more permeable.

You may not realize how many heavy metals you are exposed to every day—let alone how many are in your body right now. Approximately six million pounds of mercury are released into the environment each year. Lead, arsenic, and cadmium are present in detectable levels in our air, water, food, medicine, and industrial products. No matter where you live, you are being exposed to these heavy metals.

As with toxic mold and toxic food, some of us are more sensitive to heavy metals than others. Some people who are exposed to a small amount of mercury get very sick, for example, while others can eat high-mercury fish every day and exhibit few symptoms. But even if you're asymptomatic, the heavy metals in your body are likely preventing you from producing as much energy as you could—and therefore preventing you from kicking as much ass as you could! It's time to start doing both of those things and performing at your optimum potential.

Here are some of the most common—and damaging—heavy metals. All of them harm your mitochondria, and all of them will eventually impair cognitive function.

- **Arsenic.** Arsenic is all around us. It is used to manufacture many pesticides, herbicides, insecticides, fungicides, and rodenticides. This means it can show up in our food supply and our groundwater. Arsenic is also found in polluted water, and therefore turns up in seafood and algae. The gas contained in arsenic is used to manufacture many types of paint, enamel, glass, and metal that you may have in your home or workplace. Arsenic is also found in a place you might never suspect: brown rice. All rice contains some naturally occurring arsenic, but brown rice contains eighty times more arsenic than white rice. There, now you have permission to eat the rice that tastes good.

  In addition to being a known carcinogen, arsenic is also a neurotoxin. It reduces mitochondrial function and causes neurological problems including brain damage, nerve disease, inflammation in motor nerves, and demyelination. This will slow you down and decrease your energy production.

- **Cadmium.** Cadmium is a by-product of zinc production. For a long time, it was used to coat steel plating and to stabilize plastics, and as a pigment in glass. Recently, use of cadmium has decreased because it is so toxic, but it is also found in foods such as grains. Only three hours of exposure to cadmium causes your mitochondria to create more free radicals, become more permeable, and decrease energy production.[7]

- **Lead.** Lead is a heavy metal that was commonly used in paint until the federal government banned its use in 1978. All homes built before 1978, therefore, are likely to contain some amount of lead paint. When the paint is intact, it is unlikely to cause problems, but when the paint chips or peels, it can be easily ingested (especially by small children). If you live in a home that was built before 1978, be sure to check your windows, doors, and other areas for paint that is peeling or chipping. If it is, make sure to have it removed by a professional.

  Lead is a toxin with wide-ranging effects. In one study, rats that were exposed to low levels of lead experienced decreased mitochondrial function in every part of their brains.[8] The lead exposure also caused changes to their neurotransmitters and led to cognitive and behavioral impairments. In pigs, lead exposure has been shown to cause small structures to form in the mitochondria. These structures enlarged the mitochondria and led to mitochondrial malfunction and degeneration.[9] Human children and adults who are exposed to lead can suffer from impaired motor coordination, brain damage, seizures, convulsions, and learning and behavioral problems.

  As we've unfortunately seen in cities such as Flint, Michigan, it is also possible to be exposed to lead through water contaminated by lead pipes. If you suspect you may have lead in your drinking water, I'd recommend purchasing one of the many at-home kits available to test your tap water immediately.

- **Mercury.** As you read in chapter 6, mercury is one of the most toxic heavy metals and unfortunately also one of the most common. The presence of mercury in our seawater means that it is often found in fish and seafood, but we are also exposed to mercury in paint and some fungicides, as well as thermometers, dental fillings, and some batteries.

The first signs of mercury poisoning are fatigue, depression, sluggishness, irritability, lack of concentration, loss of memory, and headaches. Eventually, mercury poisoning can cause nerve degeneration, tremors, seizures, and permanent brain damage. In fact, the term "mad as a hatter" was first coined in seventeenth-century France when milliners, or hat makers, began experiencing tremors and pathological rage caused by chronic mercury exposure from the hat-making process.

I've experienced the effects of mercury firsthand. When I first got into yoga more than a decade ago, I had a nice little lunchtime routine of eating sushi and then doing yoga. I noticed that on days when I had sushi, my balance in certain poses was not very good. When I skipped the sushi, my balance was substantially better. To test whether or not I was actually dealing with a heavy metal, I tried taking a mercury-binding medication with the sushi—and the problem went away. It would have been easy to blame my lack of balance on just having an "off" day, but again, I believe it's critical to pay attention to the signals your body sends you. In this case, my nervous system triggered the alarm that mercury was affecting my performance.

Another common form of mercury exposure is through compact fluorescent lightbulbs. Mercury vapor is contained inside those bulbs, and when they break, they create a substantial hazard. I've taught my kids to run out of a room if they ever see a fluorescent bulb break. You should do the same thing.

PHARMACEUTICAL DRUGS

It may surprise you to learn that the FDA (Food and Drug Administration) does not require pharmaceutical companies to establish whether or not a medication approved for sale can harm your mitochondria.

In fact, a screening of more than 550 different pharmaceutical drugs revealed that 34 percent of them damage mitochondria.[10] The amount of damage that these drugs can cause depends on the dose you are prescribed as well as your particular genetics, but it is clear that a significant number of pharmaceutical drugs can have a profound impact on your performance.

Some of these medications are directly toxic to your mitochondria, meaning they impair your mitochondria's energy production. Others are indirectly toxic, meaning they increase the amount of free radicals that damage your cells while decreasing the amount of antioxidants you have to fight off those free radicals. And some medications are both directly and indirectly toxic.

Mitochondria are more sensitive to the effects of medications than any other part of your cells. When a drug makes its way into your cells, it is not evenly distributed. Instead, the mitochondria draw the drug inside of them and it accumulates there. When scientists develop drugs, they consider how it will impact various bodily systems, but they often neglect to fully consider the impact on the brain. The very medications you take to make you healthier may end up impairing your performance.

This approach seems backward to me. Let's say, for example, that the battery in your phone is having trouble holding its charge. It's having energy problems. Instead of charging the battery, you download a bunch of apps onto your phone to try to solve the problem. These apps may give you a boost of energy temporarily, but at the same time they drain the battery even more. This is what many pharmaceuticals are doing to your brain. What if you could resolve many of your medical problems by making your mitochondria stronger and more efficient at creating energy to begin with?

Don't get me wrong—I'm not suggesting that you throw out all of your prescription medications. Some of them can be lifesaving and life-enhancing for many people. But it's important to know exactly what these medications are doing to your brain so you can weigh the risks and rewards for yourself.

The following drugs will affect your mitochondria. If your doctor has prescribed any of these, I suggest having a conversation with him or her about how badly you need them and what other effects they may have.

- **Antibiotics**: Common antibiotics, including tetracycline, are now proven to cause mitochondrial dysfunction.[11] After all, your mitochondria are evolved from bacteria, and antibiotics are meant to fight bacteria! The good news is that some research indicates that the antioxidant glutathione or its precursor, cysteine, could protect mitochondria[12] when you really need to take antibiotics. This means that your body has a natural protective mechanism, but I recommend supplementing with glutathione for extra protection if you need to take antibiotics.
- **Anticonvulsants (Depakote)**: These drugs slow down the Krebs cycle so your mitochondria become less efficient at producing energy.
- **Antidepressants and Antipsychotics**: Elavil, Prozac, Cipramil, Thorazine, Prolixin, Haldol, and Risperdal all cause mitochondrial dysfunction and death.
- **Barbiturates:** Phenobarbital decreases the number and size of your mitochondria.
- **Cholesterol Medications:** Statins reduce the amount of the natural antioxidant CoQ10 in your body, which your mitochondria need to produce energy. This can lead to myopathy, a muscle tissue disease. Bile acids (cholestyramine) can inhibit the Krebs cycle but may still be worth taking in the short term because they also bind mold toxins. Another type of cholesterol medication, ciprofibrate, inhibits the Krebs cycle, making it more difficult for your mitochondria to produce energy.
- **Anti-Inflammatories:** Aspirin inhibits the Krebs cycle and causes mitochondrial uncoupling. Acetaminophen (Tylenol) increases oxidative stress, which damages the mitochondria.
- **Anti-Arrhythmics**: Amioradone inhibits mitochondrial function.

- **Antivirals (Interferon):** Treatment with interferon causes a reduction in cellular ATP levels and functional impairment of mitochondria.
- **Cancer Medications:** Doxorubicin, cisplatin, and tamoxifen all impair mitochondrial function.
- **Diabetes Medications:** Metformin makes your cells energetically inefficient.
- **Beta-Blockers:** These medications cause oxidative stress, which damages your mitochondria.

If you do have to take any of these medications, all is not lost. Following the Head Strong program will help you protect your mitochondria against these effects. Start by paying close attention to the next section to help counteract the damage from these drugs.

DETOX

Don't panic. Even if you've already been exposed to all of the toxins in this chapter, you don't have to suffer—and neither do your mitochondria. Your body has a natural detoxification system in place that is meant to process and eliminate toxins. It is an incredibly important, complex, and highly evolved biochemical system. Without it, we wouldn't be able to survive long.

One of the most interesting things about our natural detoxification system is how much it varies from person to person. When I interviewed Dr. Jeffrey Bland, the father of modern functional medicine, he told me that the detoxification system in our liver has more variability than any other system in our bodies. This means that depending on your genetics, your natural ability to detoxify pharmaceutical drugs, for example, may be far better or worse than mine. Dr. Bland explained that if you give one person with a fast detoxifying system a certain dose of a drug, the drug may have no effect at all, but if you give the same dose of the same drug to someone with a slower detoxifying system, it could kill that person.

Thankfully, there are several easy things you can do to speed up your body's natural detoxification systems. The first is to make sure that your body has the substances it needs to run these systems efficiently. When your body can't excrete toxins efficiently, they accumulate in your fat—including the fat in your brain, where they ramp up inflammation and cause neurodegeneration. When you give your body the proper nutrients and enzymes, you amplify its ability to detox. The food and supplement plan in the Head Strong program will help to ensure that you maximize your natural detoxification system.

The other main way to improve your body's ability to detox is to encourage its natural detox processes, like sweating, and to help it break down fat cells. Remember, toxins are stored in fat, so when you break down those fat cells, the toxins stored inside them are released. But once they're released from your fat cells, your liver and kidney must process them. The food and supplement plan in the Head Strong program will help support your liver and kidneys so they are effective at processing toxins.

Here are a few methods to boost your natural detox systems.

- **Sweat in a Sauna.** Sweating does more than cool you off. It also helps you get rid of a significant amount of toxins. A 2012 review of fifty studies found that sweating removes lead, cadmium, arsenic, and mercury, especially in people with high heavy metal toxicity.[13]

  Anything that makes you sweat will help you naturally detox. Exercise is one way to do it (more on that in a second), but you'll shed toxins more quickly if you use a sauna. Both traditional and infrared saunas are effective for detoxing,[14] but I prefer infrared saunas. First of all, they don't get as hot as traditional saunas. While traditional saunas heat the air around you, infrared light in an infrared sauna directly penetrates and heats your body tissue. You can stay in these saunas longer without feeling like you're going to pass out. Also, infrared light is beneficial to your mitochondria (see chapter 8 for more on this).

  Keep in mind that sweating pulls electrolytes and trace minerals from your body, so it's important to drink a lot of fluids and get plenty of salt (preferably Himalayan pink salt or another mineral-rich natural salt) if you use a sauna to detox.

- **Burn Fat with Exercise.** Exercise makes you sweat and increases lipolysis (the breakdown of fat tissue), which helps release toxins that are stored in your fat tissue. But mobilizing these toxins isn't necessarily a good thing, particularly if your body can't get rid of them. Exercise addresses the issue to a degree: it improves circulation, which provides more oxygen to your liver and kidneys so they can better filter out toxins. But if you get brain fog after exercising, you may benefit from taking a toxin-binding supplement such as activated charcoal.

- **Chelation Therapy.** If you've been exposed to a lot of heavy metals, you might want to try chelation therapy. This is the strongest way to detox heavy metals. It involves an intravenous injection of compounds called chelators that bind to toxins in the bloodstream so you can then pass them normally. Chelation therapy is effective for removing lead, mercury, aluminum, arsenic, iron, and copper. However, it can also be dangerous. If your liver and kidneys have been damaged by the heavy metal poisoning and can't process the metals, this treatment can make you very ill. Talk to your doctor before attempting this one.

- **Chlorella.** This is a type of algae that is extremely effective at binding and removing toxins from your body. It works well for detoxing from heavy metal exposure. I carry it with me and take a handful of chlorella tablets every time I eat tuna or other high-mercury fish.

Toxins are pretty scary—they're all around us, they're insidious, and they can cause us a lot of harm if we don't actively take steps to avoid them in our environment and eliminate them from our bodies. The great news is that the more Head Strong you become—with healthy, functioning mitochondria and low levels of inflammation—the more your body will be able to protect you from these brain-destroying poisons. The Head Strong program will give you a head start (get it?) toward fast detoxification, efficient energy production, and supercharged mental performance.

Head Points: Don't Forget These Three Things

- About 25 percent of us are genetically sensitive to mold and get very sick when exposed. The rest of us have symptoms that are subtle and might be written off as just a bad day.
- Mold toxins, heavy metals, and some pharmaceutical drugs are directly toxic to your mitochondria.
- Your body stores toxins in fat, so anything you do to break up the fat in your body will help you detox.

Head Start: Do These Three Things Right Now

- Check your home and office for leaks and address any potential mold issues right away.
- Keep track of how you feel in different environments—you may realize that one or more of the places you spend a lot of time in contain toxins.
- Talk to your doctor about how your prescription drugs may be affecting your mitochondria.

8

YOUR BRAIN ON LIGHT, AIR, AND COLD

Biohacking is the art of changing the environment around you and inside you so that you have full control of your body (and brain). One of the most important factors in your environment is something you probably don't pay much attention to: light. Research shows that light is a nutrient and plays a critical role in signaling mitochondria. Light tells your mitochondria what to do and when to do it, and different light frequencies send different messages. In reality, light is not just a nutrient. We've used it in medicine for one hundred years, so you could also say it's a drug.

After your brain and your heart (and your ovaries if you're a woman), your eyes contain the highest concentration of mitochondria in your body. This makes your eyes extremely sensitive to anything that might mess with your mitochondrial energy production. And some light frequencies do exactly that.

Now you might be thinking, *Why would my eyes need so many mitochondria? They're so small!* The answer is simple: it's about energy supply and demand. Your visual systems require up to 15 percent of your total energy budget.[1] Your body spends a huge amount of energy on visual processing. When you have an unstable energy supply to the mitochondria in your eyes, or just poor mitochondrial performance in general, you can suffer from brain fog and headaches and even lose your ability to perceive subtle shades of gray. In fact, changes to your perception of shades of gray (it turns out there are more than fifty of them) can be used to diagnose whether or not you have been exposed to mitochondrial toxins.

At any given moment, your eyes take in volumes of information about the world around you, and your brain requires a lot of energy to process and make sense of it all. When your eyes have to function in unnatural spectrums of light, it stresses your mitochondria, slows down your energy production, increases free radical production, and can damage mitochondria. As a result, your brain has more difficulty processing the light information your eyes take in. This can really hurt your mental performance. Mitochondria also communicate with each other,[2] so any stress to your eye mitochondria can adversely affect the mitochondria in your brain, your heart, and everywhere else.

Luckily, you're often in control of the type of light in your environment. There are many ways you can improve your mitochondrial function just by shedding some light on the problem. Literally.

JUNK LIGHT IS AS BAD AS JUNK FOOD

Today we are exposed to more unnatural-spectrum light—call it "junk light"—than ever before. Just like we accidentally screwed up our health when we began to tinker with nature and modify our food supply to create junk food, now we're screwing up our biology by modifying our natural light sources to create junk light. And your mitochondria don't survive well in it because they didn't evolve in it.

This might be the first time you're hearing about junk light, but a handful of people have been trying to get the word out about it for years. A researcher named John Ott discovered the dangers of certain light frequencies back in 1961 and has been warning us about it ever since. The bestselling author T. S. Wiley also cautioned us about the health hazards of poor-quality light almost fifteen years ago in her book *Lights Out*. I was fortunate to get a copy from her in person right after she published it, and it totally blew my mind. But for some reason, this information hasn't yet reached the mainstream.

I first became aware of light frequencies and their effect on biology even before that. As a teen, I had a pet iguana named Skippy, and I learned that he would die if I didn't expose him to the specific spectrum of light he needed. He was fine in natural sunlight, but indoors he had to have special ultraviolet reptile light. At the time, I wondered why people were so different from lizards and why light didn't matter as much to us, and I chalked it up to the fact that we're so much more evolved than lizards. It turns out we aren't so different from Skippy after all, except in our ability to *think* we are way more evolved! Light matters to us a great deal, but until the last five years we didn't know how much it mattered or that it mattered because of our mitochondria.

We evolved to absorb sunlight into our cells and our mitochondria. But in our well-intentioned quest to save energy, we've unfortunately created a blend of light frequencies for artificial light that our bodies don't recognize. We've removed infrared light, which has a wavelength that's just beyond the red end of the visible light spectrum (the electromagnetic spectrum of light that's visible to the human eye). You can't see infrared light, but you can feel it; you experience it as heat. It is the invisible part of the sun's spectrum and is necessary for most living things—including our mitochondria.

In the last thirty years, we started to completely avoid ultraviolet A (UVA) and ultraviolet B (UVB) light for the first time in history. Both of these frequencies come from the sun and have biological impacts. We now block these frequencies from our eyes through ever-present UV-filtering windows, windshields, and sunglasses, and we filter them from our skin when we use sunscreen. And that affects your whole body, because your eyes aren't the only organs that take in light. Your largest organ—your skin—also absorbs light into its cells and mitochondria. Your grandparents didn't have UV-filtering glass, didn't wear sunglasses much as kids, and didn't wear sunscreen, and they had less skin cancer and better mitochondria than we do now.

Yes, there are good reasons to filter out *some* UVA and UVB. Ultraviolet light is powerful, and we are right to be concerned about its connection to cancer. Overexposure to your skin can lead to sunburns, which can cause cancer, and overexposure to your eyes can cause permanent eye damage. It also bleaches furniture and artwork—even indoors!

It's easy to imagine that since too much UV light is bad, we should just avoid it entirely, and that's what we've largely done. But it turns out that your body requires some UV light to work properly. UVB light is vital to activate vitamin D in your body and to help set your circadian rhythm, the physiological process that tells you when to sleep and when to wake up. When I interviewed Dr. Stephanie Seneff, a senior research scientist at the Massachusetts Institute of Technology, she explained that when UVB light hits your skin, it converts vitamin D into its activated, sulfated form. Thus it's not enough to just pop a vitamin D_3 supplement. You have to activate the vitamin, and that requires exposure to real sunlight (or a high-quality UVB lamp).

Newer artificial lights like white LED (light-emitting diode) and CFL (compact fluorescent light) bulbs lack many of the sun's frequencies that our bodies and brains need. With our artificial lights, we've eliminated most of the infrared, red, and violet lighting that's found in natural sunlight, and we've amplified the blue light beyond anything we have evolved to handle (more on this in a second). We've made great strides in creating energy-efficient lighting that saves electricity, but ironically these same innovations are creating an energy crisis for our mitochondria.

This is junk light. And as I mentioned, it's exactly the same thing that happened to our food—we eliminated expensive healthy fats using the "fat is bad" myth and amplified the amount of cheap sugar beyond anything our bodies were meant to handle. You can still eat junk food, but it's not going to make you feel good. With light, we've eliminated some biologically necessary frequencies to save electricity and increased other stressful frequencies by orders of magnitude. You can still see in junk light, but you won't like the way it makes you feel.

One of the biggest problems with junk light sources is how much blue light they emit. Fluorescent lights emit substantially more blue light and less infrared light than incandescent bulbs or sunlight, which is why no one on Earth loves to be in a fluorescent-lit environment. The newer white LED lightbulbs that have invaded our cities and homes appear white, but they emit at least five times more blue light than you would find in nature, and they do it completely free of the infrared and red spectrums always found in natural sunlight.

Your mitochondria have to produce a lot of extra energy to process the blue light in LEDs, which burns oxygen and creates free radicals in the cells of your eyes. And when the mitochondria in your eyes are stressed, the rest of your mitochondria can get stressed, too, including the ones in your brain.

Recent research supports this connection between blue light and cellular damage. A 2005 study concluded that blue light "can cause cell dysfunction through the action of reactive oxygen species on DNA and that this may contribute to cellular aging, age-related pathologies, and tumorigenesis [the creation of tumors]."[3] Another study found that blue light changes mitochondrial shape and creates stress proteins in your eyes that are likely connected to macular degeneration (the deterioration of the central area of the retina, often resulting in vision loss).[4]

Macular degeneration is the leading cause of blindness in developed countries. More than a third of the population over age seventy-five has it—including my father—so I find this side effect of junk light particularly troubling. I believe that the massive and unprecedented changes to our indoor lighting and mobile phone screens will cause a huge wave of macular degeneration at younger ages than ever before. Already, several studies correlate long-term history of exposure to junk light with macular degeneration.[5, 6, 7, 8]

Industrial lighting designers simply aren't trained to consider the biological consequences of replacing streetlights and indoor lights with LEDs. They look at bulb life and electricity consumption and make the economically efficient decision to remove more expensive natural-spectrum incandescent lights, and then we pay the price biologically over time.

To be fair, processing high-quality light also generates free radicals, but there's a major difference between the by-products of processing high-quality full-spectrum light and junk light. When your eyes are exposed to high-quality full-spectrum light, the free radicals that are produced prompt the cell to produce extra antioxidants to clean up the free radicals. Your mitochondria are built to clean up their own exhaust as long as there isn't too much of it.

Blue light, however, causes an increase in free radical production but doesn't trigger the cleanup signal to increase antioxidant production. Instead of traveling to the cell nucleus, the excess free radicals stay under the cell membrane, resulting in macular degeneration and decreased energy production, and outside the eyes it can even cause premature skin aging.

Read that again. *Sitting under those bright LED and fluorescent lights at work will make you look old.*

Still want to save a couple of bucks on electricity with those "environmentally friendly" LED bulbs? Well, they're not friendly for *your* environment. We did not evolve to absorb this type of junk light. In fact, we couldn't even *see* blue until a few hundred years ago.[9] Ancient civilizations had no word for "blue." In *The Odyssey*, Homer describes the sea as "wine dark." The color blue was the last to appear in most languages, including Greek, Chinese, Japanese, and Hebrew. The evidence suggests that since they had no name for it, these people did not perceive the color blue as we know it today. Blue is a modern invention, and it is the hardest color for your brain to interpret.

So where are these damaging blue lights lurking? The main sources are the technological devices we spend our days staring into. The junk light from our smartphones, tablets, laptops, and e-readers (and from the LED bulbs in our surrounding environment) goes straight into our eyes and then into our brain, where it damages cells and lowers our performance.

Fluorescent and LED lights also cause a reduction of NAD in the mitochondria in your eyes. As you read in chapter 2, your mitochondria need NAD to complete the Krebs cycle and produce energy. A reduction in NAD causes all of the weaknesses that can stem from a lack of energy in your cells. Over time, this can change the shape of your eye, triggering nearsightedness. Even in the short term, this type of eye stress makes your brain tired. Yes, junk light can weaken your eyes so you need to wear glasses.

Light sources also regulate your circadian rhythm, the physiological process that tells you when to sleep and when to wake. Plants, animals, fungi, and even bacteria all have a twenty-four-hour circadian rhythm. Your eyes contain special light sensors that control your sleep timing. These sensors become activated at the frequency of 480 nanometers, which is in the blue spectrum. Your phone, TV, laptop, and every LED in your home all emit this light frequency. When it hits your eyes, each of the ten thousand mitochondria in your cells pays the price. Their energy production slows, they produce more free radicals, and the structure of the water they contain is altered. This causes inflammation and negatively impacts your sleep, preventing you from falling asleep easily and sleeping deeply. This stresses every system in your body and results in even more inflammation. It's a slippery, sleepy, and poorly lit slope.

Being exposed to artificial light at night also has a negative impact on your circadian rhythm. When you're exposed to daylight, your body produces serotonin, the "feel good" neurotransmitter. Your body breaks serotonin down into melatonin, a hormone that helps you sleep. If you're not exposed to enough natural sunlight during the day, you won't have enough melatonin to sleep well at night. You may have trouble falling asleep, but more likely you won't cycle through the most restful, deepest stages of sleep. Oh, and low melatonin is well known to be associated with cancer risk!

Many people think they're getting enough sleep because they go to bed at eleven and wake up at seven. That's eight hours, right? So why aren't you full of energy? Because it's the quality of your sleep that really matters, more than the quantity. Exposure to artificial light after dark slows your melatonin production even more, which prevents you from getting quality sleep and causes you to gain weight.[10] Lack of sleep and weight gain both contribute to mitochondrial inefficiency. And without sufficient energy, your brain suffers. To put it simply, junk light equals junk sleep.

There are two things you can do to limit the damaging effects of junk light, and we'll explore each in greater detail in the following two sections. The first is to cut down on your exposure to blue light. The second is to increase your exposure to high-quality light sources to balance out the excess blue you get. Sunshine is best, but you probably aren't going to spend hours outside if you have an office job or live in a Pacific Northwest rain forest like I do. Don't worry—you can have a life and feel great even if the sun isn't available as often as you need it. The Head Strong program will show you how.

RED LIGHT MEANS GO

Think about all of the different colors the sun cycles through in a day. At sunrise, it has a reddish pink glow. Toward the middle of the day it shifts to blue (with lots of UV and infrared light to balance it, which you can't see). As the afternoon bleeds into the evening, it transitions back to beautiful shades of orange and red. This light rhythm has existed longer than mammals have roamed the planet. Needless to say, we have evolved to live in accordance with this rhythm—our circadian rhythm.

The bacteria that are now our mitochondria used to float in the ocean, where they were constantly bathed in sunlight during the day. And they operate on the same circadian rhythm that we do because they're in charge! They wake during the day to eat and produce energy when food is abundant, and sleep at night to repair when it's cooler, there's no sunlight, and they have no access to food.

Your mitochondria are meant to experience red light all day, with less blue at the start and end of the day. When we spend time outside, as nature designed humans to do, our eyes (and the mitochondria they contain) are constantly exposed to full-spectrum light. Unfortunately, these days we spend relatively little time outdoors soaking up the sunshine—and our mitochondria are paying the price. When we're indoors we get tons of blue light, but no red, no infrared, and no UV. It's no wonder the mitochondria get confused and don't perform as well as they should.

Dr. Gerald Pollack, water expert and bioengineering professor from the University of Washington, has also discovered that infrared light turns the water in our bodies (and in plants) into the biologically useful EZ water that supports mitochondrial function. This EZ water should always be in plentiful supply in our cells, but the toxins we ingest from food, the environment, and junk light alter the structure of that water. This causes inflammation and leads to energy problems.

This is no small issue. Keep in mind that your body is 70 percent water. If light has the power to change that water, it also has the power to change *you*. We all intuitively know that some types of water are better than others. People feel good after doing a juice cleanse despite the sugar because they're getting so much EZ water from plants. We feel refreshed after drinking water from a young coconut for the same reason. This is also why we bother to eat cucumbers even though there's almost no actual food energy in them. These are all great sources of biologically useful EZ water, but you can also help your body make more EZ water by exposing yourself to infrared light. Doing so provides fuel for your cells and will give you dramatically more energy to be you.

For all of these reasons, exposure to natural light is essential for proper brain function. And while it's frustrating that we've limited our access to high-quality light and replaced it with junk light, it's exciting to know that there are many new technologies that can help us replicate the natural light cycle. At Bulletproof Coffee shops and at the Bulletproof Headquarters we've built in custom lighting to help employees and customers feel more energized. At home, I use my ultraviolet sun lamp for about ten minutes each morning, and I have a strip of red LED lights above my desk that I use while I work to balance out the excessive blue that's built into our technology.

If you sit in an office with junk light all day long, I highly encourage you to invest in some simple LED red lights to place in your environment, switch to halogen lights if possible, and get some quality outdoor light exposure throughout the day. Keep in mind that your skin must have direct access to this light in order for you to reap all of the benefits, so be sure to at least roll up your sleeves to get the maximum effect. Because I work from my home office, I take it a step further—unbeknownst to my team, I'm often standing naked in front of my UV-rich sun lamp while I'm on a conference call. As long as it's not a video call, it's all good!

Another way to increase your healthy light exposure is to spend time in an infrared sauna. In chapter 7, we talked about how infrared saunas can help your body detox. I've been using an infrared sauna for years, at first to detox from toxic mold and mercury exposure. Now that I know that infrared is also beneficial for my eyes and the production of EZ water, I make it a point to spend some time in my infrared sauna every day. Even once a week can make a difference.

Light and Your Skin

Collagen, the main structural protein in the human body, has a direct relationship with your mitochondria, and studies show that mutations in collagen affect mitochondria.[11] More research on this relationship is needed, but it makes sense that anything that helps you build healthy new collagen will also benefit your mitochondria.

When it comes to light, it turns out that red light is well documented to make both collagen and mitochondria grow; in fact, red light exposure causes collagen synthesis, and this is great if you want healthier skin (who doesn't?). I'm working to live to 180 years old, but I never want to look that old—so I put extra collagen protein powder in my Bulletproof Coffee to provide collagen building blocks for my body to use with my red light exposure. I also recharge my collagen and mitochondria in the REDcharger, a piece of biohacking gear with forty thousand red and infrared LEDs. I'll share some simple strategies you can use at home in the Head Strong program.

VISUAL KRYPTONITE

Limiting your exposure to junk light and increasing your exposure to high-quality light sources like red and infrared light are two easy ways to boost your energy level and mental performance. Another simple hack is to reduce the amount of visual kryptonite in your environment. Limiting your exposure to junk light is a good start, but there's more you can do. Certain types of visual stimulation such as high glare and contrast, which you might see while driving at night or when it's particularly sunny out, cause your brain to work harder than usual to process information. This stresses your brain and leads to headaches, irritability, and an inability to focus. It's also one reason you get tired in the middle of the day even if you've had a good night's sleep.

You need a relaxed brain to perform at your peak, and that means cutting down on visual kryptonite. I learned about the impact of visual kryptonite the hard way back in 2009, when I landed my dream job. I got to be an Entrepreneur in Residence at Trinity Ventures, a top-tier venture capital firm on Silicon Valley's famous Sand Hill Road. It was a job I had wanted since I was twelve years old, when I first learned what venture capitalists did. What could be more fun than identifying the next Facebook or Google when they were tiny start-ups? I came in to the office fully charged and ready to kick ass every morning.

Unfortunately, being at the bottom of the rung, I got stuck in a windowless office with poor-quality fluorescent lighting. I also had a brand-new Mac with a glossy reflective screen and an LED backlight that was much brighter than any computer I'd ever used. After just a couple of days, I started to feel profoundly fatigued, usually right at the time of day when I needed to be the most productive. I could feel my brain slowing down, and none of my normal biohacks seemed to help. The only thing that made a difference was going outside every half hour to get some sunshine.

After a month, I realized that my computer screen was making me tired. I found a company that would replace my glossy screen with an antiglare screen, and I started to get my energy back. I also adjusted the brightness and the contrast settings on my computer, installed a type of software called f.lux that controls the color of your computer screen, and wore my trademark orange sunglasses that block blue light. These small adjustments made me far more resilient throughout the day and brought my energy back. And, just once, I was mistaken for Bono.

Those orange glasses don't just make me look cool (no snarky comments, please—just let me believe this to be true), but they protect my brain. My friend Helen Irlen is a world-renowned vision educator and researcher. More than twenty years ago, she received a federal grant to study methods to help children and adults with learning disabilities. While doing so, she discovered that a large subset of the population without learning disabilities (Helen estimates 48 percent of us) has a visual processing disorder that is now known as Irlen syndrome.

If you have Irlen syndrome, you may have trouble reading for long periods of time because the contrast on the pages—in printed pages as well as e-readers—is overwhelming, and you likely get tired while driving at night because of the glare from oncoming headlights. The reason these common activities tire you out is that you use a lot of brainpower to filter out the light frequencies that your brain has trouble coping with. This puts your brain in a state of chronic stress, making it difficult for you to focus and limiting your performance. Symptoms of Irlen syndrome include headaches, eyestrain, difficulty reading, fatigue, poor depth perception, dizziness, and trouble focusing.

Irlen recommends using custom-tinted lenses that block out specific light frequencies that stress your brain. When I tested positive for Irlen syndrome and got my test results, it turned out that adding some orange, rose, and gray colors from the visual light spectrum made my brain work better. After I started wearing my own custom pair of cool orange glasses, it felt like my brain turned back on for the first time in years. I was immediately more focused and able to perform better than I ever imagined I could—even in the most distracting environments.

These glasses have made such a big difference to my performance that I wear them whenever I fly (airplanes have more junk light than almost anywhere else) or in a room with LED or fluorescent lighting that stresses my eyes. If you suspect you may have Irlen syndrome, you can find a local Irlen practitioner and get tested. I became a certified Irlen practitioner so I could test friends and clients. It's life changing when someone realizes they were pushing against their brain because of a light color mismatch.

Even if you don't have this problem, occasionally wearing sunglasses with blue-blockers indoors can help reduce eyestrain and fatigue so you can focus and perform better. Remember, about half of us have Irlen syndrome, but we don't know it. And the problem is getting worse, as blue light–heavy LED lighting is becoming the norm in offices, and we spend more time in front of computer screens than ever before. More than a few people in Hollywood sport colored glasses indoors now, and it's not just to be stylish!

THE AIR UP THERE

You've probably noticed by now that oxygen is important to your mitochondria: they need it to produce energy. This is one reason we die pretty quickly without access to oxygen. It's also why you feel like you're going to die when you try to hold your breath—your mitochondria run the show, and they want you to *really* feel the danger if oxygen is in short supply.

You obviously receive oxygen by breathing it in. The act of breathing is something that most of us take for granted, but the truth is, breathing is a unique biological function because it is the only one that is naturally both voluntary and involuntary. You don't have to think about breathing to do it; it happens automatically. But you can also intentionally alter your breath by speeding it up, slowing it down, or stopping it altogether. Breathing offers a perfect opportunity for biohackers, since one of the easiest ways to improve performance is to start by upgrading the simplest biological functions.

Think about it this way: many of the tools that transform a normal gas-burning engine into a high-performance race car engine provide mechanisms for getting more oxygen into the engine. Turbochargers and superchargers force more air into an engine using air compression so that there's enough oxygen to burn high-octane gasoline. Likewise, an important way to transform your body from an inefficient machine to a high-performance hot rod is to make more oxygen available to your mitochondria.

Unfortunately, turbocharging the human body isn't as simple as just gulping down more oxygen. Ironically, one way to increase the amount of oxygen in your body is to restrict your oxygen intake for short periods of time. This temporarily stresses your mitochondria (an example of good mitochondrial stress), which causes them to either grow stronger or die. *Eliminate the weak; train the strong* is a great algorithm for the survival of your cells.

Subjecting your body to short periods of low oxygen intake can also help it become more efficient at using oxygen when it is present. Even more interesting, short periods of low oxygen intake (known as hypoxia) increase the production of the all-important brain hormone BDNF (brain-derived neurotrophic factor), which helps support neuron growth and development.[12] Improving your body's oxygen delivery will energize your cells and make you more resilient in circumstances that limit your oxygen intake, such as travel, environmental changes, and stress. In other words, you can actually exercise your oxygen usage abilities. I do it regularly.

Boosting the body's oxygen efficiency is an important way to adapt to our changing environment. We already don't get as much oxygen just from breathing as our ancestors did because our atmosphere today contains a lot less oxygen than it did centuries ago. Even more worrying, since 2003 there has been an unprecedented drop in oxygen levels that's even more substantial than the increase in carbon dioxide.[13] You may be aware of the poor air quality when you're in a heavily trafficked city center, but recent research shows that poor air quality is also common in many indoor areas—including your gym.

A 2014 study[14] analyzed the air quality in indoor fitness centers, and the results were scary enough to make me consider the air quality before I work out in any gym. In addition to formaldehyde and other toxic compounds, the study found unacceptably high levels of carbon dioxide in the air. In a way, this makes sense. When a group of people exercises in a room without proper ventilation, the carbon dioxide they exhale gets trapped and accumulates in the environment. This often happens because building owners can save money by recycling the air instead of heating or cooling fresh air from outside, and they don't know how this impacts your biology.

The study showed that the highest levels of carbon dioxide were in an interior room used for indoor cycling. The carbon dioxide levels measured in the cycling rooms were not toxic, but they were not innocuous, either. Excess carbon dioxide in the air can make it harder to breathe and can make you feel sluggish or dizzy. Meanwhile, the more carbon dioxide is present in an environment, the less oxygen there is. All of those cyclers are breathing in the same air and competing for the same limited amount of oxygen.

This doesn't happen only at the gym. Any indoor space that holds a large group of people and has poor circulation is going to have too little oxygen and too much carbon dioxide. But you can enhance your performance in these environments by teaching your body to perform efficiently with less oxygen.

Professional athletes have known about this hack for a while—it's why many professional and Olympic athletes train in high altitudes. It causes their bodies to become more efficient and therefore higher-performing when they come back down to sea level. If this can make the difference between a gold and silver medal for an Olympian, imagine the difference it can make between your brain fog and clear, decisive thinking—if you can get those benefits without having to actually live on a mountain. Having your body consistently ready to perform at 15,000 feet provides benefits all day, every day, even if you never go above sea level.

I experienced this phenomenon firsthand when I moved to sea-level California from Albuquerque, where I grew up at an altitude of 5,000 feet. When I first arrived in California, I felt like I couldn't breathe hard while cycling because the air was so thick, but at the same time I noticed that my performance was better than usual. After about six weeks, my body adjusted and I lost my speed advantage.

We know that spending time in high altitudes will create this effect, but that's not practical for most of us. It's possible to reap the benefits of altitude training without the trouble of traveling to high altitudes, though. One way to do this is through a type of training called intermittent hypoxic training. This technique consists of intervals of breathing low-oxygen (or hypoxic) air through an air mask, alternating with intervals of breathing regular air. As your body adapts to the hypoxic air, it becomes more efficient at delivering oxygen in the blood. In addition to boosting athletic performance, this builds a tremendous amount of resilience as it pares down weak mitochondria and grows stronger ones.

At Bulletproof HQ, we use a special exercise bike that's connected to a giant bag of oxygen—but you only get to use the oxygen after you've breathed in air containing no oxygen for ninety seconds. This technique, which is called intermittent hypoxic training under load, is very effective but also expensive. (However, it's a lot less expensive than moving to Colorado!) Luckily, there's a free way to get some of the benefits of intermittent hypoxic training—and all you have to do is breathe.

Wim Hof, who holds twenty Guinness World Records for withstanding extreme temperatures, has climbed Mount Everest and Mount Kilimanjaro in only shorts and shoes. Wim is best known as "the Iceman," and you may have seen him on TV swimming among glaciers without a wet suit. He has developed a breathing technique that provides short bursts of oxygen to cells, training them to use oxygen more efficiently.

Here's how to do it: First, sit down, get comfortable, and close your eyes. Make sure you're in a position where you can freely expand your lungs. Wim suggests doing this practice right after waking up since your stomach is still empty. Warm up by inhaling deeply and drawing the breath in until you feel a slight pressure. Hold the breath for a moment before exhaling completely, pushing the air out as much as you can. Hold the exhalation for as long as you can, and then repeat this fifteen times.

Next, inhale through your nose and exhale through your mouth in short, powerful bursts, as if you're blowing up a balloon. Pull in your belly when you're exhaling and let it expand when you inhale. Do this about thirty times, using a steady pace, until you feel that your body is saturated with oxygen. You may feel light-headed or tingly, or you may experience a surge of energy that's literally electric. Try to get a sense of which parts of your body are overflowing with energy and which ones are lacking it—and where there are blockages between these two extremes. As you continue breathing, send the breath to those blockages.

When you're done, take one more big breath in, filling your lungs to maximum capacity, and then push all of the air out. Hold this for as long as you can and try to feel the oxygen spreading around your body. When you can't hold it anymore, inhale fully and feel your chest expanding. Hold it again, sending energy where your body needs it.

Bonus points if you do what Wim had me do when we demonstrated this technique onstage at our Bulletproof conference—as you are holding your lungs empty, count how many push-ups you can do before you have to breathe again. I got to twenty! It seems impossible, but you can do it, and that short bit of low oxygen forces your body to better deal with lower-oxygen environments.

I recommend you research Wim's work and watch one of his many videos online demonstrating his breathing technique. I don't think it works as well as mechanically filtering oxygen out of the air you breathe, but the Wim Hof technique is absolutely free, totally portable, and Wim is capable of things I could never do! His breathing method helps your body adapt to bursts of oxygen and puts you more in tune with the way your body uses your breath to create energy. It also makes you more resilient to cold temperatures, but there is evidence that cold temperatures themselves are good for your mitochondria.

THE BENEFITS OF BRAIN FREEZE

Cold thermogenesis is a type of cold therapy that uses cold temperatures to create heat in your body. Different types of cold therapies have been around for ages. The ancient Romans took plunges in "frigidarium baths" (large cold pools) and the Norse cracked open icy lakes for a winter swim. Even applying ice to sore muscles is a form of cold therapy. So is finishing your shower with thirty seconds of cold water!

When you soak yourself in cold water or use ice packs to lower your body temperature, your body is forced to create heat. This is called thermogenesis, a process that burns fat and stimulates the release of proteins that burn glycogen (the main storage form of glucose) from your muscles. When your muscles are depleted of glycogen, your body receives a signal to increase production of testosterone and growth hormone. This leads to a cascade of positive effects. It reduces inflammation, makes you more insulin sensitive, and stimulates autophagy so that your weak and damaged cells die and make room for new, healthy ones.

There is also evidence that cold therapy can improve thyroid and mitochondrial function. A study on rats found that cold exposure improved thyroid function,[15] while a study on humans found that it increased energy expenditure and assisted in fat loss.[16] Cold therapy also stimulates the release of the neurotransmitter norepinephrine,[17] which helps to relieve pain and signals your body to produce more antioxidants, particularly glutathione, your body's master antioxidant.[18]

A few years ago, it was all the rage in biohacker circles to sit in a tub of ice water for cold therapy. That kind of thinking earned me first-degree ice burns over 15 percent of my body (ouch!), but it turns out that it's not necessary to pack yourself with ice or for the water to be icy—cold water, about 60°F, is a great stimulus for your mitochondria. It sounds miserable, and it is for about thirty seconds. But after that, something shifts, and it feels really good.

Cold therapy also helps to tone your vagus nerve.[19] Known as the "wandering nerve" (*vagus* is Latin for "wandering"), this nerve starts at your brain stem and travels throughout the body, connecting your brain to your stomach and digestive tract, as well as your lungs, heart, spleen, intestines, liver, and kidneys. It also connects to nerves that are involved in speech, eye contact, facial expressions, and hearing.

The vagus nerve's main job is to monitor what's going on in your body and report information back to your brain. It is a key component of your parasympathetic nervous system, which is responsible for calming you down after your fight-or-flight response revs you up. The strength of your vagus nerve activity is known as your vagal tone. If you have a high vagal tone, you are able to relax more quickly after experiencing a moment of stress.

This is an incredibly important aspect of your performance. With everything you now know about how your fight-or-flight response affects your ability to focus, imagine what a difference it would make if you could recover more quickly each time your inner Labrador got riled up. Yes, the first step is to stop riling him up—which we will work to do in Part III—but it's just as important to help him calm down quickly. It's unrealistic to expect to be able to eliminate every stressor in your life. So hacking your body's response to the remaining stressors is key.

Your vagal tone impacts your performance in a number of other ways. People with a high vagal tone tend to have healthier blood glucose levels and more consistent energy.[20] People with low vagal tone are more likely to have chronic inflammation. Just as it calms your inner Labrador after it's been spooked, your vagus nerve also switches off the production of inflammatory proteins after your immune system has been activated. When you have low vagal tone, you can't shut off the inflammation as quickly, and chronic inflammation can result.

I encourage you to try cold therapy yourself, but before you do, let me offer a note of caution: safe cold thermogenesis protocols involve *gradually* increasing your exposure to cold over time. Start off by simply putting your face in cold water for a few minutes, then if you choose you can graduate to using soft-gel ice packs that won't freeze your skin, and then sitting in an ice bath for up to an hour. You have to be very careful not to overwhelm your body too quickly, or cold thermogenesis backfires. In fact, traditional Chinese medicine mostly looks down on using cold like this because it can weaken you over time.

When I tried a thermogenesis protocol, I started by plunging my face into ice water for five to ten minutes at a time. I felt great and had noticeably more energy. I was supposed to keep doing this routine for about thirty days to teach my body not to overreact to ice exposure, but I only did it for two weeks before jumping to the next step—packing ice around my body while wearing a compression shirt for thirty to forty-five minutes at a time. The shirt was meant to prevent blood from rushing in and causing bruising after the ice is removed.

I was staying in a nice hotel in New York City for a conference when I attempted this next step. It was so nice that instead of a big sink or a bathtub, my room had a shallow stainless steel sink and a walk-in shower. The place was clearly too fancy for cold thermogenesis, and I hadn't brought a compression shirt with me. But I decided to carry on with the protocol anyway. The hotel didn't have an ice machine, so I called the front desk and asked them to deliver several buckets of ice to my room.

I packed the ice into plastic ziplock bags, lay down on the bed, placed the ice across my chest, abs, and shoulders, and tried to relax and not think about the cold. After five minutes, I felt great. No shivering at all. But it was late at night, and I accidentally dozed off. About forty-five minutes later, I woke up, removed the ice, and went to bed. When I woke up the next morning, I knew right away that something was wrong. I was in pain. About 15 percent of my body felt like it had been beaten with heavy sticks.

When I looked in the mirror, I saw that every part of my body that had been covered in ice was red and puffy, as if I really had been beaten with sticks. I had left the ice packs on for far too long—to the point where I had first-degree ice burns over 15 percent of my body (according to my wife, who, thank goodness, is an ER doctor). This was not my first biohacking injury, and I'm sure it won't be my last, but you should learn from my mistake and proceed with caution. Just taking a cold shower for one minute in the morning is often enough to stimulate cold thermogenesis. So is putting your face in ice water, which I should have stuck with longer. Your body will become less inflamed, your mitochondria will get stronger, and you might even lose some weight!

Light, air, and temperature are essential components of life on planet Earth. The great thing about biohacking is that these basic elements offer opportunities to enhance performance and supercharge energy and brain power. Through safe and effective protocols that limit your exposure to junk light and increase your exposure to high-quality light, reduce visual kryptonite, enhance your oxygen efficiency, and stimulate positive cellular changes through cold therapy, you'll start getting better-quality sleep and you'll have more energy than ever before. Fat loss and better skin are the icing on the cake.

Head Points: Don't Forget These Three Things

- LED and CFL bulbs have too much blue light, which damages your mitochondria. This is junk light!
- About half of us have Irlen syndrome, which means we have trouble processing certain frequencies of light. This may be why you get tired while reading or driving at night.
- To teach your cells how to use oxygen more efficiently, temporarily expose them to less oxygen than they're used to.

Head Start: Do These Three Things Right Now

- Buy some red LED lights to balance out all of the blue light your eyes are getting from your screens.
- Wear sunglasses in an indoor environment with lots of junk light, such as an arcade or amusement park.
- Turn the water all of the way to cold for the last thirty seconds of your shower.

9

SLEEP HARDER, MEDITATE FASTER, EXERCISE LESS

It's not so much fun to learn how to become Head Strong if you're going to have to use all of your free time and energy to make it happen. The good news is that the information in this book about mitochondria not only helps us make energy more efficiently, it also tells us that there are ways to interact with our bodies that save a lot of time and energy. The Head Strong program includes biohacks that improve your mitochondria while freeing up extra time and energy every week. By working with your mitochondria instead of against them, you can improve your sleep, get more out of your meditation, and see greater results from your workouts in less time. You can do whatever you want with the extra time. And your mitochondria will thank you for it!

SLEEP HARDER

You know how you feel when you wake up from a good night's sleep, but do you know why you feel that way? Most people think it's because their bodies and minds have had a chance to rest, but that's not totally accurate. When you sleep, your body may be resting, but your brain is actually very busy. While you're off in dreamland, the brain goes into janitorial mode so it can perform for you again in the morning. Your glymphatic system, a kind of intra-brain detox system, performs the overnight maintenance work for you. Think of it as the ultimate "brain wash."

A lot of people are familiar with the lymphatic system, which uses a type of fluid called lymph to clear out toxins and cellular waste from the body. Unlike your circulatory system, which relies on a pump (your heart) to circulate blood, your lymphatic system has no pump—so it relies on muscle movement and EZ water to keep lymph flowing freely. For decades, we assumed that there was no such lymphatic system to clean the brain because the blood-brain barrier protects the brain from fluids that travel around the body. Then in 2012, researchers identified the glymphatic system, which sends clear cerebral spinal fluid through the brain's tissue, effectively flushing out cellular waste and neurotoxins from the brain and transporting them to the circulatory system. Eventually, they make their way to your liver, where they are processed as waste.[1]

Even more recently, in 2015, another group of scientists discovered that in addition to the glymphatic system, the brain also contains hidden lymph vessels. These vessels were impossible to see until imaging technology became more advanced, but now we know for sure that the brain also benefits from the lymphatic system's cleaning activities.[2] The cool thing is that because these lymph vessels eluded scientists for so long, we were able to discover the glymphatic system while we were looking for them. No one would have bothered looking for it if we knew there were normal lymph passageways in the brain. Go figure!

But it's your glymphatic system that's highly active during sleep. That's because it takes a lot of energy to circulate its cleaning fluid throughout your brain.[3] If your glymphatic system did its work during the day, you wouldn't have enough energy to hold down a job or take care of your kids at the same time. Your brain wisely waits until you are resting at night to spend energy tidying up. This is why your brain's energy production decreases very little at night even though you're not using energy to think and work and act. Your brain is instead using this energy to clean.

Interestingly, your brain cells shrink by as much as 60 percent while you sleep. This makes it easier for the fluid to circulate through your brain tissue.[4] After they've been washed clean, the cells then return to their regular size. All of this shrinking and growing and pumping (get your mind out of the gutter—I'm talking about your brain) is powered by—you guessed it—your mitochondria.

You can turbocharge your brain's maintenance system and get more cleanup done in less time if your mitochondria are working efficiently. It's a reciprocal relationship: the better your mitochondria work, the better your glymphatic system can operate, and the better-quality sleep you'll get. And the better-quality sleep you get, the better your mitochondria will work, because they'll be freshly scrubbed clean.

Every single one of the hacks I've used to boost my mitochondria has also helped me sleep better, and sleeping better has, in turn, helped to boost my mitochondria. Bottom line—if you want to kick more ass during the day, you need to get better sleep at night.

But when it comes to sleep, "better" doesn't necessarily mean more. Imagine being able to wake up feeling completely refreshed and rejuvenated after only six hours of sleep and functioning well on even less than that when life calls for it. You'd literally gain more hours in each day to get things done. That's like gaining more time to be alive. I used to aim for eight hours of sleep every night like a good boy, but I was always exhausted because I was chronically inflamed and my mitochondria weren't working well. Now I get about six hours of sleep a night (actually six hours and two minutes on average for the last 1,284 nights, according to my sleep tracking system!) and I have more energy than ever before. When it comes to sleep, quality is more important than quantity.

Sleep is important to your brain for lots of reasons. During sleep, you produce increased levels of growth hormone, which stimulates neurogenesis as well as mitochondrial growth.[5] Sleep also strengthens connections between brain cells and improves your memory by allowing your brain to process experiences and solidify new memories. In a 2014 study, researchers looked at the impact of sleep on new learning in mice. They taught a group of mice a simple task and allowed them to practice it for an hour. Then the mice were separated into two subgroups. One group slept for seven hours, and the other group was forced to stay awake. The mice that were allowed to sleep had significant dendrite growth in their brains (remember, dendrites are the rootlike structures that allow for information to be passed between neurons). The mice that did not sleep exhibited less new dendrite growth.[6] Those new dendrites created pathways where new skills became embedded in the mice's brains. Without them, it would be much more difficult for the mice to access the new skills and information they'd learned.

Good-quality sleep also affects energy levels by helping to keep your blood sugar stable. Getting consistently poor-quality sleep causes a 40 percent decrease in blood sugar regulation.[7] In other words, when you don't sleep well for an extended period of time, your body eventually becomes insulin resistant and less efficient at making and using energy. You've probably experienced this firsthand in the form of carb cravings and mood swings after pulling an all-nighter or simply having a bad night of sleep. People who don't sleep well are also at an increased risk of weight gain and obesity. As we know, this leads to a ripple effect of dangerous conditions, including chronic inflammation, which slows down energy production in your mitochondria.[8]

Everything goes back to mitochondria—even snoring. Studies show that when mitochondria are not working at their best, many common sleep disorders, including sleep apnea,[9] can result. And sleep disorders may pose real threats to your health. People who snore are at nearly double the risk of developing diabetes, obesity, and high blood pressure as compared to those who sleep well. If these people wake up feeling groggy or have trouble falling asleep, their risk for these diseases goes up 70–80 percent.[10] Is this because their bad sleep causes poor blood sugar regulation or because they have a mitochondrial dysfunction that causes them to get bad sleep, which then causes poor blood sugar regulation? I would bet on both.

Every aspect of the Head Strong program is designed to fire up your mitochondria so that you can benefit from more restorative sleep, stable blood sugar levels, an increased rate of neurogenesis, and more energy. And remember, all of these benefits are the result of improving the *quality* of your sleep. With better sleep, you can sleep for fewer hours than you do now and have significantly more energy. We'll discuss lots of ways to help improve your sleep, including getting the right light exposure and adjusting your nighttime habits to help facilitate a good night's rest. But perhaps the best thing you can do to start sleeping better right now is to manage your stress.

MEDITATION FOR CALMER, HAPPIER, AND SHAPELIER MITOCHONDRIA

By now, you understand why the Head Strong program is set up to help eliminate many of the physiological sources of stress to your brain, like toxic foods, environmental toxins, junk light, and visual kryptonite. In some ways, those are the easy types of stressors to handle. Unfortunately, psychological stress is not as straightforward—and it can really mess with your sleep. We all have different stress levels, but the best remedy I know for any type of mental or emotional stress is good old-fashioned meditation. Or at least, a totally modern, biohacked version of it.

Many people are skeptical about meditating, but meditation is not just another woo-woo trend. Researchers release new studies seemingly every day that prove just how beneficial meditation is for your brain. Since I started working in Silicon Valley, I've seen a dramatic shift in the way people approach meditation. Twenty years ago, few executives would admit to having a meditation practice. Today, it's such a highly regarded tool for stress management that it's hard to find someone who's willing to admit that he or she *doesn't* meditate. Many high-powered executives, such as Arianna Huffington, have shared publicly that they use meditation to improve their sleep or boost their performance. Google even provides meditation classes to thousands of employees because they believe it's worth the effort.

The science of meditation may be new, but meditation itself is not. People have been practicing meditation in cultural and religious contexts for thousands of years. But it doesn't matter what religion (if any) you ascribe to—anyone can practice meditation without any religion whatsoever. In fact, if everyone meditated for just ten minutes a day, we'd all be healthier, happier, and a lot nicer to one another. The reason is that meditation drives self-awareness. And when you have more energy, it's easier to pay attention to yourself, your thoughts, and your actions. With awareness comes control.

Studies show that meditation changes the brain on a structural level.[11] Think of it like strength training; when you lift weights, you gain visible results in the form of stronger, shapelier muscles. A regular meditation practice also yields visible results—you develop more folds in the outer layer of the brain, a trait that's highly correlated with intelligence across species.[12] When you have more folds in your brain, it's easier to process information because your neurons can access more surface area within the same skull volume. This allows them to communicate with each other more quickly and efficiently. Aging naturally flattens these folds, but meditation slows this process.[13]

Meditation also helps to thicken areas of the cortex and insula—regions of the brain associated with complex thought, bodily awareness, concentration, and problem solving—that also typically thin with age.[14] And meditation has been shown to significantly reduce levels of the stress hormones cortisol and adrenaline, effectively reducing inflammation and calming your inner Labrador so that you can maintain focus and emotional stability under even the most challenging circumstances. Basically, meditation can help you feel sharp, stay calm, learn new things, and stop being a jerk. It's no surprise, then, that research shows that meditation can improve your relationships and make it easier for you to achieve your life goals.[15]

Of course, we wouldn't be talking about meditation if it didn't also have a positive impact on your mitochondria. A 2013 Harvard Medical School study found that people who practiced just twenty minutes of "relaxation response"—which included diaphragmatic breathing, body scan, mantra repetition, and mindfulness meditation—experienced several health benefits including reductions in hypertension, reductions in infertility, and reductions in depression. These benefits were attributed to "improving mitochondrial energy production and utilization and thus promoting mitochondrial resiliency." The effect was stronger in people who meditated regularly than in the newbies, but a noticeable difference was observed even after just one meditation session. It's almost as if those little ancient bacteria that run our cells are listening to your meditation and changing in response!

While studies have proven that meditation has a measurable impact on the brain, scientists are still investigating the mechanisms of how meditation improves your biology in other ways. But we do know that our mitochondria are involved in regulating every bodily system that is impacted by meditation. Is it too much of a stretch to suggest that our mitochondria, which sense the environment around us and tailor their energy production accordingly, also respond to meditation, as the earlier study essentially implied? To me, this is the most likely explanation as to why meditation has so many positive effects across bodily systems, and it is one that conveniently sidesteps the spiritual implications of meditation that make so many people uncomfortable. You might take issue with the idea of connecting to the divine, but it's hard to argue against making your mitochondria work better by taking control of your stress levels. And there's nothing stopping you from doing both!

Of course, if meditation were easy, everyone would already have a stress-free life. And various religious believers wouldn't have spent thousands of years sitting in caves and monasteries learning how to do it. While meditation may look simple, believe me, it takes a lot of time and effort to sit in silence and calm your mind on a daily basis. I've practiced many forms of meditation throughout the years, and here's why.

When I was a kid, I thought I'd be happy when I made $1 million, so I put all my effort into that. I ended up making $6 million when I was twenty-six years old. You know what I told myself then? That I'd be happy when I made $10 million, so I kept pushing, which caused me to lose it all when I was twenty-eight years old. And the whole time, I wasn't happy. I was stressed, and my biology paid the price.

It was then that I realized that no amount of thinking and rationalizing could compute my way to happiness, and I couldn't buy my way to it, so I resolved to hack the problem. After all, what other tools did I have besides hacking?

I scraped together what little money I had left, and I did the following: I traveled to Tibet to learn meditation from the masters who originated it; I attended a ten-day silent Buddhist meditation retreat in Nepal; I spent five years practicing a daily Hindu-inspired Art of Living breathing meditation with a group of uber-successful Silicon Valley friends; I tried a traditional ayahuasca ceremony in the Peruvian jungle long before tech CEOs could book them online; I learned advanced yoga and pranayama; I fasted for days alone in a cave in the Sedona desert; and I learned to hook electrodes to my head to measure my meditating brain waves. None of these experiences were easy, but they were beneficial and informative.

Across all of these different forms of meditation, one common denominator I noticed was that I experienced better results when I received some sort of external feedback. Think of it like working out alone versus exercising with a trainer. Most people progress more efficiently when they have someone coaching and pushing them. The same thing goes for building new "muscles" in your brain—or mitochondria, as the case may be.

So, am I suggesting you start shopping for a guru? Not exactly. Even the most observant meditation teacher can't tell you exactly what you're doing right or wrong, and certainly not in the moment—any feedback would be delayed by the few seconds it takes them to observe your behavior and tell you about it. From the perspective of your nervous system or mitochondria, a few seconds might as well be a hundred years.

That is why I believe the best and most efficient way to meditate is in conjunction with technology that can provide instantaneous feedback. The most effective tool I have found is called EEG (electroencephalogram) neurofeedback, although *any* technology that provides fast feedback can improve your meditation. With neurofeedback, a practitioner applies sensors to defined points on your scalp. These sensors monitor your brain waves and send them to a computer. (Brain waves are electrical pulses from masses of neurons communicating with each other, and they can be detected easily from your scalp.) The computer then converts your brain waves into sounds or images that offer you a visual or audible representation of what's happening inside your brain as you meditate.

Every thought, feeling, or emotion you experience influences your brain waves, and they fluctuate constantly. When your brain waves are out of whack, you can have all sorts of emotional and neurological problems ranging from ADHD to anxiety and depression to rage and bipolar disorder. On the flip side, those conditions can cause your brain waves to be out of whack, too. Either way, properly constructed neurofeedback offers a rapid-fire look inside those brain waves, picking up a signal thousands of times every second. The results, charted on-screen in real time, help you learn to perfect your meditation in far less time than it normally takes, and even experienced meditators can gain new deep levels of awareness in shockingly small periods of time. Meditating with the aid of neurofeedback is like following a two-lane illuminated highway versus trying to follow a winding path in the dark with no flashlight. Both roads will get you to your destination, but one requires a lot more time because progress is slow and you can get lost on the way.

To try neurofeedback, you can buy your own unit, find a practitioner, or go to a focused clinic. There are tens of thousands of neurofeedback practitioners and hundreds of different types of technologies. They range from profoundly effective (in the hands of a proficient neurofeedback expert, you can gain enormous mental benefits) to profoundly dangerous (in the hands of an inexperienced or ill-informed practitioner, some people have gained anxiety and lost restful sleep for months). Just as with any other health care decision, it's vital to do your research and choose a reliable practitioner.

The big risk with neurofeedback is that many systems are based on a model that collects data from the brain waves of hundreds of people to calculate the average response of a "normal brain." Then the technology trains your brain to that standard. If you have a below-average brain, this type of training can give you your life back. However, if you have an awesome brain, this type of training can take away some of your awesomeness. It's just like high school. A class taught to the lowest common denominator can make an F student into a C student. But at the same time, that class might not challenge an A student, who would be held back from achieving his or her potential.

I started experimenting with neurofeedback in 1997, when I purchased my first machine to use at home. Along the way, I trained at many clinics, and my experience at one of them did have a negative impact on my cognitive abilities. It took me a full two weeks in my own private neurofeedback facility to reverse those effects and feel like myself again. These technologies are always evolving. On the Bulletproof website, I maintain an updated list of several *safe* and affordable systems you can purchase to use at home. These range from $300 entry-level systems with limited power to $5,000 systems that approach clinical-grade capabilities.

If you're looking to take it to the next level, you can graduate from meditating in the traditional sense, where you try to clear your mind of all thoughts, and use technology to help you focus on certain types of beneficial states. Hemoencephalography (HEG), which I use at home, is a type of feedback that focuses on increasing blood flow to the prefrontal cortex, the "human brain." To do this, you strap a sensor to your forehead and then concentrate or think happy thoughts. When you do it right, blood rushes to the frontal lobe, the sensor detects the change in blood flow, and it provides feedback. With practice, you can start to grow the blood flow to your prefrontal cortex so that you're using your "human brain" more often than not—even when you get stressed out. This type of neurofeedback is particularly effective for people who suffer from ADHD.

But by far the most extreme (and most effective) form of neurofeedback I've ever done is called 40 Years of Zen. I've done this program with more than a hundred clients over the past five years. It's offered only in one place on earth, a $2.5 million facility in Seattle that looks something like Xavier's School for Gifted Youngsters from the *X-Men* movies, but without the tennis court and invisible jet. It is an intensive, five-day, all-day, all-in program designed to put you in the same state of mind as an advanced Zen practitioner who's been practicing meditation for decades.

Instead of relying on technology to tell your brain what to do, this program combines meditation techniques to show you what your mind is already doing so you can change it. The core is a Neurofeedback Augmented Reset™ technique that shows you how your brain is automatically and unconsciously responding to events in your life and then shows you how to turn off the reactivity. Another technique, called Retroframing™, works with feedback to allow clients to create the automatic subconscious reactions you want, all guided by neurofeedback. There are other modules to increase voltage potential in the brain (obviously driven by mitochondria) and to increase the speed of synaptic firing. All of this was unimaginable even ten years ago, because signal processing hadn't come this far, and our neurological knowledge of advanced Zen meditation states wasn't there yet, either.

This program has changed my cognitive abilities so much that I've spent ten weeks of my life—about seventy days—with electrodes stuck to my head, training my brain and the mitochondria in my neurons until they wouldn't take any more. I wouldn't be writing this book or running a fast-growing company today if it weren't for the benefits I've gained from this training. I'm such a believer that I recently funded the creation of new 40 Years of Zen custom-built hardware and software so I can continue to push my brain to new levels. Nothing else has ever changed me this profoundly, and the effects are quantifiable. The height of my brain waves (amplitude) is about four times higher than it was when I started neurofeedback, and my brain waves are measurably more organized and efficient.

Even people who are already highly functioning experienced meditators find this form of neurofeedback to be hugely beneficial. Vishen Lakhiani, author of the *New York Times* bestseller *The Code of the Extraordinary Mind* and CEO of Mindvalley, the largest meditation website in the world, participated in this training with me. He's been teaching meditation for twenty-five years and he said that it was the single most effective form of meditation he'd ever done. His book, *Code of the Extraordinary Mind*, contains a chapter describing his experiences with this type of neurofeedback.

Of course, the downside of having a dedicated team of technicians, coaches, and neuroscientists train your brain using three different types of equipment for a week is that it's quite expensive. I am working to make this kind of intensive brain training more widely available—I'd like to see it used in every high school in the country. Giving our young people the ability to enhance their minds in this way could reduce a lot of suffering and quickly and dramatically change our society for the better. In fact, I have an initiative under way to make this type of brain training available for students globally at an affordable price.

Obviously, the 40 Years of Zen program is an extreme form of neurofeedback that isn't yet accessible to everyone. You are hearing about it only as an example of what's possible when you combine the power of meditation techniques with neurofeedback technology at the highest possible levels. But even very affordable, simple forms of feedback from your nervous system can offer positive benefits. One such program is called heart rate variability (HRV) training. All it requires in the way of technology is a smartphone and a heart rate sensor that you can buy online, and it's so simple to practice that I taught my kids how to do it when they were four years old!

In HRV training, you start by taking slow, deep breaths guided by an app on your phone. Then you do something in your chest that feels funny, and when you do it right, the app rewards you with a green light and a chime. What you are doing is changing the spacing between your heartbeats to activate your parasympathetic nervous system so you can take control of your body's stress response. It would probably take you a couple of years of meditation practice to learn how to do this without technology. But with a sensor and immediate feedback, you can learn this skill in just a few weeks.

The effects of being able to control your stress response and calm your body's fight-or-flight hormones are profound. If you practice HRV regularly, you'll have more energy and less stress. I use this technology when coaching executives, and I use it before I go onstage to give a keynote address!

If all this technology is just too much for you, don't sweat it. The Head Strong program includes a zero-tech, completely free mitochondrial meditation created by one of twelve living masters of an ancient form of Chinese energy medicine.

EXERCISE YOUR BRAIN

While it's important to reduce your psychological stress so you can sleep and perform better, stressing your cells through exercise is one of the best things you can do to make them stronger. We've known for a long time that exercise promotes mitochondrial health, but we're still learning about all of the exciting connections between exercise and mitophagy (killing weak mitochondria), neurogenesis (growing new neurons), and mitogenesis (growing new mitochondria).

Exercise is one of the best ways to stimulate the release of an important protein called PGC-1 alpha (peroxisome proliferator-activated receptor-gamma coactivator-1 alpha), which helps to regulate metabolism and mitogenesis. (Cold exposure also stimulates PCG-1 alpha production—this is how cold thermogenesis helps create new mitochondria.)[16] But just like with sleep, the quality of your exercise matters more than the quantity—simply logging in a few extra minutes on the treadmill won't cut it. It takes high-intensity interval training (HIIT) to release this protein.[17] We'll talk more about the benefits of HIIT in just a bit.

When you exercise, your muscles also release a protein called FNDC5 (fibronectin type III domain-containing protein 5). Part of this protein goes into the bloodstream and increases levels of brain-derived neurotrophic factor (BDNF) in the hippocampus, where neurogenesis occurs. As you read earlier, BDNF is a protein that supports the growth and differentiation of new neurons and is one of the most important substances in your body for neurogenesis. In 2008, Dr. John J. Ratey, a professor at Harvard University, first referred to BDNF as "Miracle-Gro for the brain." When researchers treat neurons with BDNF in a lab, the neurons spontaneously grow new dendrites that aid in learning.[18] It is one of my primary targets for biohacking because BDNF stimulates neurogenesis, neuroprotection, neuroregeneration, cell survival, synaptic plasticity, and the formation and retention of new memories.[19]

Researchers have known about the link between BDNF and exercise for a while, but it wasn't until 2013 that they discovered a link between PCG-1 alpha and BDNF. It turns out that increasing PCG-1 alpha raises FNDC5 production, which leads to an even greater increase in BDNF.[20] It makes sense that the birth of new neurons and new mitochondria would be connected, but it's amazing to know that something as simple and available to all of us as exercise can help build new brain cells *and* the mitochondria needed to power them.

Researchers at Northwestern University have found yet another exciting connection between exercise and brain performance. Their work reveals that exercise lowers the activity of bone-morphogenetic protein (BMP), a protein that reduces your rate of neurogenesis. At the same time, exercise raises your levels of noggin (I swear that's the real name), a protein that counteracts BMP and actually increases your rate of neurogenesis.[21]

Exercise not only helps you become fitter, it also encourages the survival of your fittest mitochondria. That's because exercise lowers the mTOR protein, which helps your body weed out the weak, dysfunctional, or mutated cells, and either kills them off or makes them stronger. Only the strong survive! Without damaged mitochondria slowing things down and with your previously weak mitochondria now working better, your energy production improves dramatically. Studies have shown that having fitter mitochondria also helps to reduce your chances of developing many neurodegenerative diseases and has even been shown to have a neuroprotective effect in Parkinson's patients.[22]

You probably already know that regular exercise helps to lower your blood sugar levels and make you more sensitive to insulin. This not only helps you to stay trim, it also keeps your energy levels stable and increases endorphins, "feel good" neurotransmitters, which can help fight depression. In fact, researchers have found that regular exercise is *at least* as powerful as antidepressants in fighting depression.[23] Finally, exercise also improves circulation, which reduces inflammation and allows more oxygen to make its way into all of your tissues, including your brain. This helps your mitochondria to make energy more quickly. And with more blood flow to your liver, you also eliminate neurotoxins more easily.

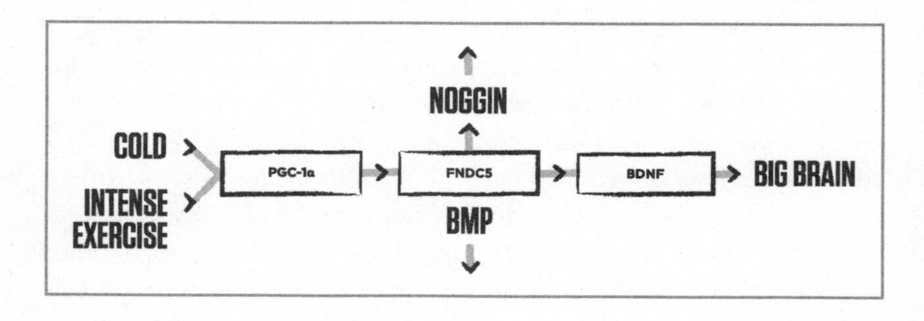

New neurons, new mitochondria, fewer toxins, and fewer dysfunctional mitochondria. Oh, and more stable energy and less depression. Are you ready to hit the gym? I thought so. But remember, not all forms of exercise are created equal. Different types of exercise actually give you slightly different mental (and physical) benefits, so it's important to create a balanced exercise plan to get the most out of your workout. Here's a quick breakdown of which forms of exercise are best for your brain.

FUNCTIONAL MOVEMENT

You don't have to go to the gym or pay for a fancy cycling class to benefit from exercise. All you have to do is move. Activities like taking a walk, practicing yoga, going for a hike, bike riding, jumping rope, or running around with your kids on the playground will all improve your brain and stimulate neurogenesis.

Research has shown that daily movement is better for your brain than less frequent bouts of exercise.[24] This is why it's so bad for your brain to remain seated at your desk all day long. If you work in an office *and* drive to work, it's essential to find some time during the day to get up and move around. We're talking about movement that is far below what you'd consider exercise—just walking around is fine.

In fact, walking is particularly good for the brain. In one study, a group of healthy adults who took forty-minute walks three times a week for a year experienced a measurable increase in the size of their hippocampus.[25] Since this is the part of the brain where neurogenesis occurs, we can assume that walking increased the participants' rates of neurogenesis.

Yoga is another form of movement that boosts brainpower. Researchers at the University of Illinois found that a single twenty-minute session of hatha yoga significantly improved participants' results on a working memory test they were administered after the yoga session. The study indicated that yoga helped the participants stay focused and take in, retain, and use new information. You might assume this to be true for any form of exercise, but in fact, participants performed significantly better immediately after the yoga practice as compared to their results after moderate to vigorous aerobic exercise for the same duration of time.[26] In another study, elderly patients who practiced yoga experienced growth in their hippocampus.[27] Yoga has also been shown to help reduce stress, which lowers inflammation throughout your body, including in your brain.

Another way that yoga benefits us is that it incorporates cross-lateral movements, or movements in which one of your limbs crosses the center line of your body. These movements increase blood flow to all parts of the brain as well as the number of synaptic connections in the brain.[28] Cross-lateral movements also force the left and right hemispheres of the brain to work together and are being studied as a possible treatment for dyslexia.[29]

RESISTANCE TRAINING

As the name suggests, resistance training involves pushing against a force that resists movement. Some examples of resistance training include weight lifting, kettle bell training, and body weight workouts. Resistance training is typically brief and intense, and that short burst of stress is great for your body and your brain.

It's no surprise that resistance training makes your muscles stronger, but it also improves your brain so you can move *better*. In a recent study, fifteen men lifted weights for fourteen weeks. At the end of the fourteen weeks their muscles could generate more force, of course, but what's more interesting is that their neural drive—their ability to send electrical signals from their brains to their muscles—became stronger and faster.[30] With greater neural drive comes more precise control of the way you move.

Resistance training also aids the body's natural detoxification process. Earlier, you read about your lymphatic system, which relies on physical movement to pump a liquid called lymph around the body and collect waste from your cells. One study found that ten to fifteen minutes of brief muscle contractions increased lymph flow by 300–600 percent.[31] This is probably at least in part because moving all of that water around inside your cells as you exercise creates more EZ water, which itself helps mitochondria and lymph flow.

Perhaps most important, resistance training has profound mental health benefits that can have a major impact on your performance. An excellent 2010 review of randomized, controlled studies showed that strength training significantly decreases anxiety, improves memory and cognition, reduces fatigue, and makes you happier. All of this is due to the boost in endorphins and the increased rate of neurogenesis, mitogenesis, and mitophagy that resistance training provides.[32] Resistance training also causes a sharp increase in testosterone and a 200–700 percent increase in human growth hormone to help you grow new neurons.[33]

The biggest hurdle when it comes to resistance training is that most people tend to either not do it at all or overdo it completely. Remember, this type of training stresses your muscles and your mitochondria. That's a good thing, but you can't stress them every day—they need time to recover before you stress them again. It's during rest that your body incorporates the changes in your brain that stem from exercise. In other words, it's during rest that your new neurons and mitochondria are born and your old, dysfunctional mitochondria either get stronger or die.

Our bodies are already overly stressed from sources like toxic food, environmental toxins, and junk light. Exercise momentarily weakens you to make you stronger in the long term. But if you stress yourself too much by overexercising, you can make yourself too weak to reap the benefits. Overtraining is a major problem today, and one of the first things that happens when you work out too much is that your cortisol levels rise. When cortisol goes up, BDNF drops.[34]

Dr. Doug McGuff, the author of *Body by Science* (a book I highly recommend), suggests doing resistance training only once every seven to ten days. His research shows that this offers greater benefits than more frequent exercise, and of course it saves you a lot of time. But the catch is, you have to work *hard* when you are training. You smack down your mitochondria with intensity and then back off and let them recover.

During the two-week Head Strong program, you'll do resistance training once a week. That's it. You'll do plenty of movement on other days, but this one hard-core workout each week will maximize the benefits for your brain.

ENDURANCE TRAINING

Endurance training, commonly referred to as either cardio or aerobic exercise, relies on oxygen from your lungs to produce energy. Anything that makes you breathe heavily, such as running, biking, or swimming, stresses your aerobic system and tests your endurance, bringing a host of benefits to the table.

While both resistance and endurance training increase BDNF, resistance training only does so for a short period of time directly after you exercise.[35] Endurance training seems to boost BDNF levels more permanently. A randomized, controlled study of twelve men showed that three months of daily cycling nearly quadrupled their resting BDNF levels.[36]

Endurance training is one of the best ways to get your body to release endorphins[37] and has been shown to have a positive effect on mood. This explains the "runner's high" that you feel after an endurance workout. You don't have to run until you're blue in the face, though. One study revealed that just ten days of power walking was enough to significantly lessen symptoms of depression.[38] Other research has shown that creativity peaks after a good endurance workout.[39]

Most people think that they have to do thirty to forty minutes of cardio exercise three or four times a week to get results. That's a huge time commitment when you're busy kicking ass and getting stuff done. And it certainly doesn't leave a lot of time for other forms of exercise, like resistance training. The good news is, there's a type of exercise that combines all of the benefits of resistance and endurance training to give you the best of both worlds.

HIGH-INTENSITY INTERVAL TRAINING (HIIT)

HIIT workouts alternate between intense bursts of strenuous exercise and brief periods of active rest. You might sprint for sixty seconds, walk for thirty, do push-ups for sixty, walk for thirty, and so on. The icing on the cake is that HIIT is efficient. You don't need to spend an hour on the treadmill followed by an hour in the weight room. With HIIT you can get an amazing workout in a very short amount of time.

HIIT subjects your muscles and cardiovascular system to extreme stress and then allows them to recover during the "active rest" period. This active rest period keeps your heart rate elevated so that you can still reap the rewards of aerobic exercise. It's the happy marriage of two schools of fitness thought, and it might be more effective than either resistance or aerobic exercise alone. In fact, studies have shown that HIIT is up to ten times more effective at increasing growth hormone[40] than resistance training or endurance training.[41]

Interval training is also good for your heart. One of the best measures of cardiac fitness is your ejection fraction, which is the amount of blood your heart can pump in a single beat. Unfortunately, most medium-intensity cardio workouts actually *decrease* your ejection fraction. The best way to increase it is through running intervals such as 400-meter sprints. The normal protocol is to run as fast as you can for 400 meters, walk for a minute as you recover, and then repeat the process until you can't run anymore.

There is one upgrade to HIIT that only saw the light of day at the 2016 Bulletproof Conference—until now. The secret is to sit—or better yet, lie on your back—for a full ninety seconds between sprints. This allows your nervous system to come back into equilibrium more quickly, so you reap more of the benefits from the sprint. It sounds strange, but you can actually feel the difference this makes when you're done working out. Surprisingly, the source of this upgrade is John Gray, the famous author of *Men Are from Mars, Women Are from Venus*, who discovered it while doing research on nutrition and exercise for one of his books and shared it onstage.

When you do this simple and quick workout during the two-week Head Strong program, you'll experience a huge boost in your cardiovascular health and your mental performance.

EXERCISE AND WATER

When you think about water in relation to exercise, your mind probably goes straight to the role of hydration. And sure, staying hydrated is important, but even more important, I would argue, is the profound impact of exercise on the water in your cells.

You know how essential EZ water is for your mitochondria. Shaking water molecules—a natural result of most forms of exercise—

creates EZ water. But it's not only your mitochondria that benefit from this "shaken" water. *All* cell membranes are made up of small droplets of fat suspended in water. When these tiny drops of fat are flexed and shaken, it creates a piezoelectric effect, which allows materials to generate an electric charge in response to stress.[42, 43] This electric effect creates EZ water in your cells and allows them to do their jobs more efficiently.

While any movement shakes up the water in your cells to some degree, high-vibrational movements work best. I use two pieces of equipment to take advantage of this effect (and keep my inflammation-derived love handles and man-boobs as small as possible). One is the Bulletproof Vibe, which vibrates the entire body thirty times a second at a frequency that NASA discovered helps astronauts recover after spending extended periods in space.

The Bulletproof Vibe also helps drain lymph from your body. When you ditch the toxins, your mitochondria work better, and so does your head. Standing on the Vibe for a few minutes gives you a very large amount of vibration without as much effort as it would take to do it the way your mother did, with an old-fashioned rebounder or trampoline. But jumping on a mini trampoline is a good way to get that piezoelectric effect in your cell membranes and cause lymphatic drainage of toxins that slow down your brain. You can get a rebounder or mini trampoline on Craigslist for about $10.

Another way to generate this effect is to jump rope or do jumping jacks, though both of these activities require more cardiovascular endurance than using a rebounder. When you get more bounce for less effort, your EZ water gets more benefit in less time before you have a chance to get tired out. While you can't see the microscopic changes to the water in your cells, you will notice that your muffin top magically and quickly shrinks.

The other piece of technology I use is called Atmospheric Cell Trainer (ACT)™, which requires a large piece of equipment that is basically the cockpit and canopy from a fighter jet. You sit in the cockpit while a giant turbine changes the air pressure to go from Mount Everest heights (22,000 feet) down to sea level and back. It does this quickly, which causes all of the cells in your body to expand and contract all at once, creating a massive piezoelectric effect, more than you can get from any workout. This type of cellular "exercise" is also believed to increase the number of stem cells in the body.

To try ACT you're going to need to find a high-end sports training facility outfitted with the necessary equipment. But if that's not possible, there is a relatively affordable hack you can do instead. Changes in atmospheric pressure have a name—they're called sound waves. I've long been interested in "sound therapy" devices (the kind that go on your body, not in your ears) and have seen some impressive results. This is an area of biohacking with lots of innovation and not enough scientific validation, but it's worth paying attention to.

Perhaps the vibration from sound helps to structure the water in our cells. I'm not sure, but I do have a sound therapy system installed in my infrared sauna. It has "bass shaker" sound transducers mounted to the seats, so when you play sound therapy tracks, you feel it in your bones. You can get these for home theater use for a couple hundred dollars, but the best option is to use a SubPac device, which transfers sound waves directly to your body, on the back of your desk chair or in a wearable vest.

I realize that recommending sound therapy is going out on a limb—we know that vibration makes more EZ water, but there is no concrete evidence that can confirm that sound therapy moves EZ water into your cells. But I've seen it help people, and I think it's a hack worth considering, especially if you don't use a rebounder or trampoline. I used a SubPac while writing this book!

Sticky Water Versus EZ Water

We're still learning more about the way water works in our cells. In 2015 scientists in Germany discovered that free radicals make the water inside your mitochondria more viscous and sticky and that exposure to UV light can undo that effect. Imagine a boat trying to get through water that's thick like glue. That's the challenge faced by your mitochondria when the water inside is sticky. It's harder for them to produce energy, making them less efficient.

This concept is so groundbreaking that the scientists involved in this study said they expect their work to "have broad implications in all fields of medicine."[44] I do, too.

Their work shows how important it is to help your body create the right structure of water so you can have better mitochondrial function. That means reducing the number of free radicals your mitochondria create. If you're inflamed all of the time, your mitochondria make lots of free radicals. You can think of them as exhaust from poorly running mitochondria. Your cells' water will get sticky, which makes it harder for you to make energy in your cells. And that makes for more free radicals . . . which make for less energy . . . and pretty soon you're stuck with brain fog, muscle pain, and a head that's anything but strong.

Laser and light therapy can fix the sticky water problem by reducing the creation of free radicals and inflammation, and certain foods and supplements can do the same things. The combination can be transformative; it was for me.

MELANIN AND WATER

New research shows that exercise (and light!) affects your brain in ways we never previously imagined. Scientists in Mexico looking at eye diseases were trying to figure out how some parts of the eye have huge amounts of oxygen deep inside them—more than can theoretically come from breathing. They eventually turned their focus to melanin, the pigment that makes our skin dark and that is also in our eyes and our brains. The researchers observed that when it is exposed to sunlight or mechanical vibration, melanin has the power to break water apart, freeing up oxygen and electrons for your mitochondria to use to make energy.[45] This is very new science that most people have never heard of, and more research is needed—but it has exciting implications.

It turns out that melanin is incredibly important for our mitochondria. And guess where we get melanin? We make it by linking polyphenols together! This means that the more polyphenols you eat, the more melanin you can make, and the more oxygen and electrons your mitochondria will have access to—if you get adequate sunlight and use exercise to shake the water in your cells.

Of course, talking about polyphenols means we have to talk about America's number one polyphenol source, coffee. In addition to polyphenols, coffee contains melanin and similar compounds called melanoids. This new science about EZ water and melanin might explain an observation that has kept me wondering for years. EZ water forms when you shake or blend water, and it forms more easily when there are small droplets of fat present because fat helps create a piezoelectric effect. This makes me think that there is a lot more to Bulletproof Coffee than I ever realized.

When I created the recipe, I could not explain why it was so important to actually blend (not just stir) the coffee with the butter and Brain Octane Oil. I also could not explain why this combination worked well with coffee and with other less caffeinated dark beverages like chocolate or real vanilla beans, but it did not work as well with lighter beverages. And finally, I could never explain why you can't get the same effects by drinking a cup of black coffee and eating a stick of butter with a shot of Brain Octane Oil. If you believe the researchers working on melanin and EZ water (and I do), a possible answer to this puzzle is that the act of blending the fat and water mechanically creates EZ water. Blending that EZ water with the polyphenols, melanin, and melanoids[46] in the coffee could have a direct effect on the coffee by creating free oxygen and electrons—before it even enters your body.

In Tibet, I noticed that nomads with very little money—people who carry all of their belongings on the backs of yaks—had blenders hooked to portable batteries, just to make yak butter tea. It made no sense until I realized that blending is part of what makes it work!

At the time of this writing, I'm in discussions with Dr. Gerald Pollack to see if we can measure these effects to prove the hypothesis. Whether it is true or not, there must be a reason that drinking a cup of black coffee and then eating a stick of butter with some Brain Octane Oil does not do the same things for your brain as blending them all together. I believe the reason, like the answers to so many things, lies in your mitochondria.

Now that you know all of the levers and triggers that you need to pull in order to maximize your brain performance, it's time to put them into action! The Head Strong program will not just teach you how to eat to fuel the mitochondria in your brain, detox from environmental toxins that limit energy, and maximize your light exposure, breathing, workouts, and sleep, but it will also help you stack these up so that you gain the most benefit in the shortest amount of time. The next two weeks are going to be nothing short of life changing. Are you ready? Let's get to it.

Head Points: Don't Forget These Three Things

- When it comes to sleep, focus on quality more than quantity.
- Meditation changes your brain on a structural level—for the better.
- You need lots of time to recover in between intense bouts of exercise—at least several days.

Head Start: Do These Three Things Right Now

- Jump on a trampoline or do jumping jacks to shake the water in your cells and make them more EZ.
- Get some extra sleep tonight to give your brain a chance to form pathways between neurons and solidify new memories. Later on, you'll learn how to get better-quality sleep.
- Try breathing in for five seconds, holding for five seconds, breathing out for five seconds, and holding the out breath for five seconds. Do it five times in a row.

PART III

THE TWO-WEEK HEAD STRONG PROGRAM

By this point you know how much control you have over your mitochondria, the little critters that power your brain—and that's got to feel pretty exciting. So many of the environmental, diet, and lifestyle factors that drive them—light exposure, breathing, temperature therapy, exercise, sleep habits, and, of course, the foods you eat—are completely up to you. You do thousands of things every day that affect the way your brain creates energy. Choosing to do even just a few of them better will lead to powerful change.

But as Spider-Man's uncle Ben said, "With great power comes great responsibility." And just as you can choose to do things that boost your mitochondria, you can choose to do things that slow down your energy production and make you tired, forgetful, unfocused, and moody. In fact, you're probably already doing some of those things without realizing it. Choosing to keep doing those things will prevent you from living up to your full potential as a human being and giving as much as you can to your community, your family, and our world.

When you make the choice to ramp up your energy *and* stop doing things that slow you down, you can finally have all of the power it takes to be you. It feels amazing—physically, emotionally, and even spiritually—to know that you're making the most of your time on this planet and contributing as much as possible to the things you care about. And that is what you have to look forward to starting within the next two weeks.

Throughout the two-week Head Strong program, you are going to eat to fuel the mitochondria in your brain and begin simple habits that will have a profound effect on how you feel and perform every day. Once your brain reaches its fullest potential, it's up to you to decide what you'll do with all of that mental energy and power. I know you'll choose wisely.

10

EAT TO FUEL YOUR BRAIN

Over the next two weeks, you are going to eat the healthiest, most brain-boosting foods on the planet. During this time, you can get yourself into a state where your body makes its own ketones, which will shift your mitochondria into high gear. Typically, it takes a very careful near-zero carb diet and at least four days to get yourself to this state. But the recipes proposed here help you achieve ketosis with Brain Octane Oil instead. As you read earlier, Brain Octane Oil is 5–6 percent of what you find in coconut oil distilled into a concentrated form to provide a more powerful energy source. If coconut oil is like a weak beer, then Brain Octane Oil is strong vodka or Everclear. It is a source of ketones from outside your body that keeps you from having to make them yourself.

If you don't want to use Brain Octane Oil, you can follow a standard high-fat, low-carb template. To do so, eliminate the carbohydrates and reduce the protein by 50 percent in the following recipes. To get the full effect, you want at least a few ketones in your blood most of the time. You can get keto strips to measure your urine, and they should be at least a very light pink. The minimum blood level you want every day is 0.5.

Anyone can get into ketosis for a little while by avoiding all carbs; doing it for long periods of time is harder, which is why I developed Brain Octane Oil—it amplifies ketosis even if carbohydrates are present. But all of the recipes in the Head Strong program use specific ingredients that raise your ketones higher than you can with normal food. And while you can't do that with just coconut oil, you should feel free to use coconut oil on top of the oil in these recipes (as long as you like it).

You can mix and match the recipes in this chapter to create daily menus that suit your lifestyle. Just pick one meal from each category (breakfast, lunch, dinner, and an optional dessert and snack). No need to count calories or fat grams or stress out about whether or not you're eating too much. Listen to your body. If you're still hungry, eat more. If you're full, stop eating. Remember that calories are actually used to measure energy—and if your body and your brain are going to run at full power, they're going to need plenty of calories. The high content of healthy fat in these recipes will fuel your brain and keep you satiated.

A couple of things to keep in mind: First, if you are accustomed to eating a diet that is high in sugar or processed foods, you might go through a brief period of withdrawal as you drop those foods and start eating nutritious whole foods in their place. Don't panic if you feel tired for a couple of days when you start the program. That's just your body fighting to give up the substances it has grown addicted to. Rest assured that this will pass quickly and will be followed by a stronger, clearer mind and body. And happily, supplements that improve mitochondria usually help with this kind of tiredness in the meantime.

It's also important to note that if you have been on a low-fat diet, it may take your body a little time to adjust to the increase in high-energy fats that comes with the Head Strong diet plan, particularly the potent Brain Octane Oil. If you use its less powerful cousin, generic MCT oil, instead, you won't get as many ketones and there is a pretty big chance you'll end up with what you might call "disaster pants." This is something you'll want to avoid, so be careful with your quantities from the start. The amount of Brain Octane Oil that I call for in these recipes is a general guideline for beginners—I use a lot more at home. Start slowly, see how your body responds, and work your way up from there.

Before starting the program, I recommend dedicating a shopping trip or two to source high-quality ingredients. A local farm or farmer's market is definitely your best bet for finding great produce and grass-fed animal products. If that is not an option for you, try your best to source organic, local produce (always make an effort to find organic produce, even when it's not specified in the recipes) and grass-fed meat from grocery stores in your area. I actually like buying a lot of my groceries online, where it can be easier to find exactly what I'm looking for. Remember, we're going for local and organic so that your mitochondria will be exposed to fewer toxins.

To help you get started, here is a shopping list for the next two weeks:

Proteins

- Gelatin (try Bulletproof CollaGelatin for two times the protein)
- Pastured Eggs
- Pastured Bacon (don't skimp on quality!)
- Ground Grass-Fed Lamb
- Ground Grass-Fed Beef or Pastured Pork
- Pastured Lamb Leg or Shoulder
- Ground Bison or Grass-Fed Beef
- Wild Salmon Fillets
- Wild Alaskan or Sockeye Salmon, Cold Smoked
- Wild Sea Scallops

Fats:

- Brain Octane Oil
- Grass-Fed Ghee (Bulletproof makes one that is 100 percent butterfat from grass-fed cows)
- Grass-Fed Butter

- Coconut Milk, Full-Fat, BPA-free and guar gum–free
- Coconut Crème
- Olive Oil
- Avocados

Herbs and Spices

- Sea Salt (or, preferably, pink Himalayan salt)
- Xylitol from hardwood, or Stevia
- Vanilla Extract (Bulletproof VanillaMax is lab tested to be mold-free)
- Sage
- Fennel Seeds
- Cayenne Powder
- Oregano
- Rosemary
- Mint
- Cilantro
- Basil
- Fresh Ginger
- Shallots
- Cumin
- Coriander
- Thai Basil
- Turmeric

Veggies

- Cauliflower
- Broccoli
- Red Bell Peppers
- Leeks
- Asparagus
- Carrots
- Broccoli Sprouts
- Iceberg Lettuce
- Zucchini
- Cucumbers
- Romaine Lettuce
- Celery
- Thai Chilies
- Lemongrass

Fruits

- Frozen Blueberries (high-quality organic brands to reduce chance of mold)
- Shredded Coconut
- Tomatoes
- Lemons
- Limes
- Blackberries
- Raspberries

Nuts

- Pistachios
- Almond Butter
- Pistachio Butter

Miscellaneous

- Upgraded Bulletproof Coffee (your mileage may vary with other high-quality coffee beans—organic is not enough because even organic coffee beans may contain mold, but try for single-estate, washed coffee from high elevations in Central America to minimize the risk of mold)
- Upgraded Collagen from Grass-Fed Cows
- Matcha Green Tea Powder
- White Rice
- Gluten-Free, Grain-Free Bread or Crackers (I like Mary's Gone Crackers, even though they contain some gluten-free grain)
- Chocolate/Cocoa Powder (Bulletproof makes a lab-tested one)
- 85% Dark Chocolate
- Apple Cider Vinegar

BREAKFAST

Choose one of the following meals for breakfast each day on the Head Strong program.

LIQUID ENERGY BREAKFAST: BULLETPROOF COFFEE WITH UPGRADED COLLAGEN

SERVES 1

Who says a breakfast that gives you steady, all-day energy and sharpens your thinking has to include solid food? This one will keep your motor (and your gray matter) purring till lunch. It raises your ketones so you can start enjoying 147 electrons per molecule instead of the measly 36 you were getting from your old smoothie breakfast.

INGREDIENTS

1 to 2 cups Upgraded Bulletproof Coffee

1 to 2 tablespoons unsalted grass-fed butter

1 to 2 tablespoons Brain Octane Oil

1 tablespoon or more Upgraded Collagen Protein

Stevia or xylitol (optional)

1. Brew the coffee using a metal filter, such as a French press or a gold filter drip.
2. Pour the coffee into a preheated blender (preheat the blender by filling it with hot water for a few minutes and then dumping the water out). Add the butter, oil, and collagen protein. Blend until a layer of froth forms at the top, at least 20 seconds. Add stevia to taste, if using.

If you find yourself feeling tired or having sugar cravings a few hours later, the most likely culprit is mold in your coffee beans. Try

Bulletproof beans or use the guidelines mentioned to find the beans that are most likely to be mold-free.

BRAIN SUNRISE

SERVES 1

This is a relatively fast and easy breakfast for any day of the week. You'll get ketones for your mitochondria from the Brain Octane Oil and lots of healthy fats from the avocado, bacon, and eggs for your cell membranes and myelin. These fats will keep you satiated and focused at least until lunch, and the eggs will help your body make the important neurotransmitter acetylcholine so you can sleep well at night and wake up refreshed for another beautiful brain sunrise tomorrow.

INGREDIENTS

2 or 3 strips pastured bacon

1 or 2 eggs from pastured chickens

2 tablespoons apple cider vinegar

½ organic Hass avocado

1 tablespoon Brain Octane oil

Himalayan salt and fresh herbs to taste

1. Preheat the oven to 320°F.
2. Bake the bacon in a roasting pan or a rimmed baking sheet for about 10 minutes, turning once. Conserve the delicious bacon fat. (If you cook the bacon on a stovetop, low and slow is the way to go—no smoking, sizzling, damaged-fat bacon!)
3. Poach the eggs in water that has been combined with the apple cider vinegar. (Hint: swirl the water before cracking the eggs so the eggs stay in the center of the pan.)
4. Slice the avocado lengthwise into thin crescents, arrange the

crescents in an arc across the plate, and then place the bacon around the rim. Arrange the poached eggs at the base, like the sun peeking up over the horizon.

5 Mix 1 tablespoon of the conserved bacon fat with the oil. Drizzle this energy ambrosia over your Brain Sunrise. (Save any leftover bacon fat and store in the fridge for use in future dishes.) Garnish with some salt and fresh herbs, and serve with black coffee, green tea, or no-sugar hot chocolate for polyphenols. That's what I call a good morning!

SMOKED SALMON SCRAMBLED EGGS WITH SHAVED BROCCOLI

SERVES 2 OR 3

This meal is chock-full of healthy fats, particularly omega-3s from the salmon. The salmon will also help you make acetylcholine, GABA, dopamine, and serotonin, while the eggs will assist even further in the production of acetylcholine. This breakfast also supplies plenty of polyphenols from the broccoli, and the lemon gives it a tart kick in the ass so you can spend your day kicking more ass.

INGREDIENTS

1 head of organic broccoli

Juice of 1 lemon

2 tablespoons Brain Octane Oil

6 pastured eggs

1 tablespoon grass-fed ghee or coconut oil

Sea salt

⅛ teaspoon dill (optional)

4 slices of smoked wild salmon (look for wild Alaskan or sockeye)

1 Cut off the whole florets from the head of broccoli. Using a vegetable peeler, peel off the tough outer layer of the broccoli stalks. Then

use a mandolin or a peeler to cut very thin slices of broccoli (if they curl, that's okay). Imagine that they're veggie fettuccine. Slice each floret in half. Place the broccoli (florets and stem slices) in an adjustable steam pot cage and steam in a saucepan with a small amount of water until it's bright green. Remove the broccoli from the heat and rinse it with cold water in a colander to preserve the blush of true al dente vegetables. Arrange the shaved broccoli in the center of a plate as a base for the eggs, with the florets on the side.

2. Combine the lemon juice and the oil in a small bowl. Set aside.
3. Crack the eggs in a separate bowl. Place a stainless steel or cast-iron pan over medium heat. Add the ghee. Whisk together the eggs until just combined, add the salt and dill, if you like, and then pour them quickly into the pan. Stir continuously and cook until the eggs are cooked to your liking.
4. Spoon the eggs onto the broccoli stalks, and then drape the salmon over the eggs. Then quickly drizzle the salmon, eggs, and florets with the lemon–Brain Octane Oil mixture. To mix it up, replace the broccoli with steamed asparagus.

BLUEBERRY UN-CHEESECAKE

SERVES 4

When a dessert is this good for your brain, don't hesitate: eat it for breakfast!

If you speak Italian (or haute cuisine), you might call this a *panna cotta*. For me it has all of the creamy richness of cheesecake with none of the nutritional nightmares that normally accompany that dish. Plus, the blueberries offer brain-boosting polyphenols and the avocado, butter, coconut milk, and Brain Octane Oil offer plenty of healthy fat. If you prefer, you can obviously eat this one for dessert, instead!

INGREDIENTS

1 cup fresh or frozen organic blueberries

4 cups full-fat coconut milk, BPA-free and guar gum–free

Up to 4 tablespoons xylitol or stevia

2 tablespoons Bulletproof CollaGelatin or 1 tablespoon gelatin

2 teaspoons vanilla extract

4 tablespoons unsalted grass-fed butter

1 tablespoon Brain Octane Oil

2 cups shredded coconut

½ organic Hass avocado (optional)

1. Place the berries in a deep-sided dish. Heat 1 cup of the coconut milk, the xylitol or stevia, and the gelatin in a saucepan over medium heat until the gelatin is dissolved.
2. Place the remaining 3 cups of coconut milk in a blender with the vanilla, butter, and oil. Blend thoroughly and then add the hot coconut milk/gelatin mixture and shredded coconut. Pulse the blender until mixed. (If desired, add the avocado and pulse some more.) Pour the entire blender contents over the blueberries and place the dish in the fridge for an hour to set. Add more berries to the top to finish.
3. In a hurry? Freeze for just 15 minutes to get the jelled consistency, and use frozen berries instead of thawed ones.
4. To turn this recipe into a snack that will rock your next dinner party or get your kids clamoring for brain-boosting foods, just fill small paper cups with blueberries, add the mixture to them, and slide them in the fridge. These turn into the healthiest and most delicious "gelatin shots" I've ever had.

LUNCH

Choose one of the following meals for lunch each day.

GRASS-FED BEEF OR PASTURED PORK AND ROAST-VEGGIE FEAST

SERVES 4

Imagine you're a medieval king, chowing down on delicious wild boar and organic veggies that were cooked on a spit over an open fire. (Okay, medieval people *only* had organic veggies.) That may seem difficult to re-create, but this brain-boosting meal smells just as good and is a hell of a lot easier to prepare. The pork will help you make acetylcholine and the veggies supply tons of polyphenols. Cook the food low and slow, so that you don't create cooking toxins, and the result will be mouthwatering and powerfully energizing.

INGREDIENTS

1 basket of organic tomatoes—about 12 ounces

2 medium organic leeks, sliced in ¼-inch rings

1 bunch of organic asparagus, about 1 pound

Grass-fed ghee

2 pounds ground pastured pork or beef

½ teaspoon fennel seeds

½ red bell pepper, minced

1 teaspoon sage

1½ teaspoons sea salt, plus more as needed

2 to 4 tablespoons Brain Octane Oil

1. Preheat the oven to 350°F.
2. Chop the tomatoes and leeks and break off the bottom ends of the asparagus. Coat a roasting pan (with high sides, to avoid losing

anything over the side) with ghee, put in the vegetables, add more ghee, and cook them in the oven for 20 minutes. While the veggies roast, mix the pork with the fennel seeds, bell pepper, sage, and salt. Shape the meat into patties. Coat another roasting pan with ghee.

3 After 20 minutes, reduce the oven temperature to 320°F and leave the veggies in there. Put the patties in the oven on another rack to roast for 35 to 45 minutes (depending on how thick you make the patties).

4 Roast until the meat is thoroughly cooked. Remove the veggies when they are turning golden around the edges. (Trust your nose. When roast veggies are perfect, they smell divine.) Add salt to taste. For ketones and extra flavor, sprinkle everything with Brain Octane Oil just before serving. For variety, you can substitute fresh fennel and Brussels sprouts for two of the roasted vegetables above.

LAMB BURGERS WITH CUCUMBER GUACAMOLE

SERVES 4 TO 6

This decadent meal is full of healthy fats, polyphenols, and even EZ water from the cucumber in the guacamole. Meanwhile, the lamb meat will help your body make acetylcholine, GABA, and serotonin. And the Brain Octane Oil changes the creaminess of the guacamole and will add an amazing mouth feel, flavor, and those coveted ketones. Expect to feel calm, satisfied, and focused after this meal. Who knew a burger could make you so Head Strong?

BURGER INGREDIENTS

1 head organic iceberg lettuce

3 large organic carrots

2 or 3 organic yellow or green zucchini

2 teaspoons dried oregano

1 teaspoon dried rosemary

2 teaspoons turmeric

Sea salt

2 pounds ground lamb

Grass-fed ghee

½ cup organic broccoli sprouts

BULLETPROOF GUACAMOLE INGREDIENTS

4 ripe organic Hass avocados, pitted and scooped out

2 to 4 tablespoons Brain Octane Oil

2 teaspoons sea salt

1 to 3 teaspoons apple cider vinegar or lime juice

Pinch of ascorbic acid (keeps it green for a long time!)

½ English cucumber or a whole Persian cucumber (peeled)

¼ cup chopped cilantro or herbs of your choosing

1. Gently peel off the iceberg lettuce leaves to make top and bottom "burger buns." Use a spiralizer to turn the carrots into springy corkscrew threads. Or, if you don't have time, just grate 'em. Slice the zucchini into thick matchsticks and place the matchsticks in a steamer. Wait to cook them until the burgers are almost done.
2. *For the burgers:* In a large bowl, mix the oregano, rosemary, turmeric, and salt into the lamb. Form eight burger patties. Gently cook the patties in ghee in a covered cast-iron skillet over medium heat, for about 4 minutes on each side. (By avoiding high heat, you avoid degrading the lamb while it's cooking. The goal is to cook it through without burning or caramelizing the meat, as that creates heterocyclic amines and AGEs, which break your mitochondria.) Reserve the pan juices.
3. Steam the matchstick zucchini to al dente. Be vigilant. It won't take long.
4. *For the guacamole:* Place the avocado oil, salt, vinegar, ascorbic acid, and cucumber in the bowl of a food processer or blender and blend until very creamy. Stir in the cilantro or herbs of your choice (20 percent of people think cilantro tastes like soap; it's genetic).
5. *To finish:* Place the burger patties on lettuce leaves and dollop a healthy amount of the guacamole on top of the burger. Top with broc-

coli sprouts and spiraled carrots and finish with another lettuce leaf. Balance the plate with the zucchini. As a final touch, spoon the juice from the skillet (there should be a fair amount) on top of the zucchini. That juice has some powerful brain-boosting fat in it.

GREEN MIND, CLEAR MIND MATCHA BOWL

SERVES 1

This nontraditional meal is great for those days when you're so busy bringing it that you can't really deal with cooking, or when you're still full from your Head Strong breakfast but need a quick burst of brain energy to power you through the rest of the day. This contains high amounts of polyphenols and far more collagen than bone broth!

INGREDIENTS

1 ripe organic Hass avocado

1 organic Persian cucumber or ½ organic English cucumber (peeled)

½ cup coconut cream

1 teaspoon Matcha green tea powder

1 to 2 tablespoons Brain Octane Oil

5 mint leaves

2 tablespoons Bulletproof Collagen

Stevia

¼ cup pistachios or as many as you want!

Shredded coconut

Handful of mint sprigs, for garnish

1. Place the avocado, cucumber, coconut cream, tea powder, oil, and mint leaves in a blender and blend until well combined. Next, add the collagen. Just blend it enough to mix it in and no more. (If you over-blend, it can degrade the collagen.) Then add stevia to taste and stir.

2 Pour the mixture into a bowl. Sprinkle the pistachios and shredded coconut liberally on top. Have as many pistachios as you like, as long they are not old or discolored (pistachios are at high risk of mold formation). They are great sources of polyphenols. Garnish with mint sprigs. Eat with a small teaspoon to prolong the pleasure of texture and color and taste. Make it a mindfulness exercise to pay attention to the distinct and steady lift in energy you feel as you eat.

DINNER

Each day, choose one of the following meals.

GRILLED CILANTRO-LIME WILD SALMON AND FRESH BLACKBERRIES WITH BRAIN RICE

SERVES 4

This is one of the most powerful brain-boosting meals on the planet. You'll get omega-3s from the salmon, polyphenols from the berries, and just enough carbs from the low-carb cooked rice to keep your energy up all night while you're sleeping so your brain can run its cleaning system. The salmon will also help you make acetylcholine, GABA, dopamine, and serotonin, plus it has DHA and EPA fatty acids. Whew! And on top of all that, it's delicious.

SALMON INGREDIENTS

2 tablespoons high-quality olive oil

Juice of 1 lime

1 tablespoon Brain Octane Oil

2 tablespoons chopped organic cilantro

Sea salt

4 skin-on wild salmon fillets (8 ounces each), skin scored lightly—you can also use one large 2-pound fillet

BRAIN RICE INGREDIENTS

2 to 3 cups uncooked white rice

1 to 2 tablespoons Brain Octane Oil

4 tablespoons unsalted grass-fed butter or grass-fed ghee

1 head of organic broccoli, cut into spears

Handful of chopped fresh organic basil (chop right before using)

Handful of pistachios or as many as you want!

1 cup fresh organic blackberries

Sea salt

1 lemon, cut into wedges

1. *For the Salmon:* Heat your grill to medium-high. Mix together the olive oil, lime juice, Brain Octane Oil, and cilantro. Add sea salt to taste. Rub the mixture into the salmon. Reduce the heat to medium-low to avoid charring the skin. Place the salmon on the grill surface, skin-side down, and cook for 6 to 12 minutes (depending on thickness), until the fish is medium-rare.
2. *For the Brain Rice:* Steam the rice as you normally would, but also add the oil at the start of cooking. It adds ketone power and reduces the carbohydrates your body can absorb from the rice. It also creates prebiotics from the starch in the rice!
3. Heat 2 tablespoons of butter in a medium saucepan over medium-low heat. Spoon in the steamed rice and stir. Add salt to taste. Cook for 1 to 5 minutes, stirring frequently, until hot. Stir in the remaining 2 tablespoons of butter and cook for 1 minute more. Cover to keep warm as you grill the salmon.
4. *To Finish:* Finely chop 2 tablespoons of raw broccoli and set aside. Steam the rest of the broccoli. Use the rice as a bed for the cooked broccoli. Top with the basil, the raw broccoli (for enzymes), and the pistachios. Plate the salmon with the blackberries and lemon wedges and serve the finished rice alongside it.

HACK-YOUR-OWN TACOS

SERVES 4 TO 6

These delicious tacos contain a lot of polyphenols and antioxidants that will clean up the free radicals in your mitochondria, giving you a noticeable lift in brain energy. My kids love having this Head Strong version of taco night—and you will, too.

INGREDIENTS

2 pounds ground bison or grass-fed beef

2 tablespoons unsalted grass-fed butter or ghee

Juice of 1 lime

1 to 3 teaspoons cayenne powder

1 teaspoon oregano

1 teaspoon ground cumin

Sea salt

3 or 4 organic carrots

16 organic romaine lettuce leaves

1 cup organic broccoli sprouts or another kind of sprout if you prefer

½ bunch of organic cilantro

Bulletproof Guacamole (see page 219)

1. Sauté the meat in a large, heavy-bottomed pan over medium-low heat until cooked gently but thoroughly. Your goal is not to brown the meat. Drain the liquid and save it for another cooking adventure. (Bison juice has lots of flavor and good fats. It lasts for a few days in the refrigerator.) Add the butter, lime juice, cayenne, oregano, cumin, and salt and stir thoroughly over low heat just until the flavors combine.
2. Grate the carrots into a bowl and wash and dry the individual romaine leaves. Set out the carrots and lettuce banquet style in the middle of the table with the sprouts, the cilantro, and a generous bowl of guacamole. Then hand everyone a plate, and let them hack their own lettuce-shelled tacos.

COCONUT-LAMB CURRY, THAI STYLE

SERVES 2 TO 4

This is a very flavorful and versatile meal that provides lots of polyphenols, antioxidants, and good fats. The lamb will also help you make acetylcholine, GABA, and serotonin to ensure a good night's sleep after eating this for dinner.

CURRY PASTE

2 to 5 fresh Thai chilies (also called bird's eye chili), stems removed (2 will add mild heat; 5 will kick your ass)

1 stalk of lemongrass (just the white part)

1-inch piece of fresh ginger, peeled and sliced

1 shallot, diced

1 teaspoon ground cumin

1 teaspoon ground coriander

1 red bell pepper

Juice of 1 lime

1 small bunch of cilantro

Handful of Thai basil leaves—you can use regular basil if needed

BASE AND MEAT

2 tablespoons grass-fed ghee

1 pound boneless pastured lamb leg or shoulder, cut into 1½-inch pieces

1 15-ounce can full-fat coconut milk, BPA-free and guar gum–free

1 teaspoon Brain Octane Oil

½ cup free-range chicken stock or vegetable stock

3 or 4 stalks of organic broccoli, cut into medium-size pieces

½ small or medium head of organic cauliflower, cut into medium-size pieces

Himalayan pink salt or sea salt

2 cups Brain Rice (optional; see page 222)

1. *For the Curry Paste:* Put all of the curry paste ingredients in the bowl of a food processor or blender and blend until smooth. If you use a small blender, you can add some of the coconut milk to make it blend more easily.
2. *For the Base and Meat:* Melt 1 tablespoon ghee in a large saucepan. Add the lamb and lightly cook until it just starts to change color at the edges. Remove the meat and set aside. Add the curry paste to the same pan (the lamb will already have given a wonderful base of flavor to the pan and the ghee). Add the remaining tablespoon of ghee and cook over medium heat for 3 to 4 minutes. Add half of the coconut milk and simmer uncovered for 10 minutes. The curry should begin to thicken.
3. While the sauce simmers, heat the oil over medium heat in a separate large saucepan. Add the lightly cooked lamb, turning the pieces to brown on all sides but taking care not to char them, and cook until medium-rare, 7 to 8 minutes. Once the lamb is ready, add it to the curry pan along with the remaining coconut milk, chicken stock, broccoli, and cauliflower. Simmer for another 3 to 4 minutes. Salt to taste. Serve the curry over Brain Rice for a carb day, or serve it straight up with veggies for a low-carb day.

BACON-GINGER SCALLOPS WITH CAULI RICE

SERVES 2 TO 4

This dish has everything your brain needs—omega-3s, polyphenols, antioxidants, and lots and lots of yummy fats. It's fine to replace the cauli rice with white rice if you feel like you need more carbs. As long as you add Brain Octane Oil, you'll still be getting ketones!

BACON-GINGER SCALLOPS INGREDIENTS

3 tablespoons grass-fed butter or ghee

10 stalks of lemongrass (8 of them you will use as big toothpicks)

1-inch piece of fresh ginger, peeled and finely grated

1 pound wild sea scallops, tendons removed and patted dry (no need to rinse)

8 pieces of thin-cut bacon

1 teaspoon ground turmeric

CAULI RICE INGREDIENTS

1 head of organic cauliflower

2 tablespoons unsalted grass-fed butter

2 tablespoons Brain Octane Oil

Sea salt

SALAD INGREDIENTS

1 head of organic romaine lettuce

1 cup broccoli sprouts

4 stalks of organic celery, sliced

1. *For the Bacon-Ginger Scallops:* Preheat the oven to 320°F. Put the butter, 2 stalks of chopped lemongrass (the white part), and the ginger in a medium saucepan. Heat over low heat, stirring often, for 20 to 30 minutes, until the flavors have infused. Make sure the mixture does not boil! Once infused, remove the saucepan from the heat. Place the scallops in a small bowl. Pour the mixture over the scallops. Wrap a piece of bacon around each scallop and secure each with one of the lemongrass stalks. Set the scallops on a rimmed baking sheet, sprinkle with turmeric, and place in the oven. Bake for 8 to 15 minutes, checking often, until the bacon is crispy. The cooking time will depend on the size of the scallops. Remove the lemongrass toothpicks before serving, or face the wrath of your diners.
2. *For the Cauli Rice:* Grate the cauliflower or use a food processor to pulse it to the right texture so that it resembles rice. Heat a large sauté pan over medium heat and melt the butter. When the butter melts, add the riced cauliflower. Don't be afraid to crowd the pan, as it will aid in the cooking process by creating a steamer effect. Caution: you don't want to brown the cauliflower. Cook it gently for 5 to 10 minutes, stirring and turning over often. A Dutch oven can make this process less messy, as it has high sides that help prevent the cauli rice from being tossed overboard. Once the cauliflower is cooked through, turn off the heat, add the oil, and season with salt to taste.

3. Serve with the scallops and a romaine lettuce side salad that's topped with broccoli sprouts and sliced celery.

SNACKS

The meals in the Head Strong program are satisfying and satiating, and you'll likely find that you won't need many snacks. But when you do want something to munch on between meals, there are a few easy go-to options. If you feel yourself flagging midday, have one of these snacks for a quick boost in mental energy.

- 1 ounce 85% dark chocolate
- Small serving of Blueberry Un-Cheesecake (page 215)
- Head Strong Quick Bites: Place one or more of the following toppings on a piece of grain-free bread or grain-free crackers such as Mary's Gone Crackers: 1 tablespoon grass-fed butter; a slice of wild smoked salmon; ¼ avocado, mashed.

DESSERT

Yes, you can eat all of that delicious fat for breakfast, lunch, and dinner, and you still get to have dessert, too. While these desserts are optional, I do recommend having at least a small serving of one of these recipes before bed to make sure your brain has enough energy to run your glymphatic system while you sleep. When was the last time you thought of eating dessert as doing something good for your brain? This is your new normal.

BRAIN SHAKE

SERVES 1

This creamy, high-fat smoothie is loaded with polyphenols and will give you plenty of energy to sleep well!

INGREDIENTS

2½ cups full-fat coconut milk, BPA-free and guar gum–free

2 tablespoons almond butter or pistachio butter, if you can find it, for extra polyphenols

2 tablespoons Brain Octane Oil

1 small organic Hass avocado

½ cup raspberries or 3 tablespoons Bulletproof Hot Chocolate mix

2 scoops Bulletproof Collagen Protein

½ teaspoon ground Ceylon cinnamon

1 Unfair Advantage ampule for mitochondrial boost (optional)

Stevia or xylitol

Ice (optional)

Blend! Pour! Drink! (Don't you love easy recipes?)

3-BERRY GELATO

SERVES 4

This dessert fights inflammation while delivering huge amounts of antioxidants.

INGREDIENTS

6 ounces organic blueberries

6 ounces organic raspberries

6 ounces organic blackberries

1 15-ounce can full-fat coconut milk, BPA-free and guar gum–free (well shaken)

3 tablespoons Brain Octane Oil

2 large pastured egg yolks

⅛ teaspoon vanilla extract

Up to 3 tablespoons xylitol (optional)

Up to 2 grams ascorbic acid (see note) for tartness (optional)

1. In a blender, combine the blueberries, raspberries, blackberries, coconut milk, oil, egg yolks, and vanilla. Blend until smooth. At this point, take a moment to taste the mixture before you add the xylitol and/or the ascorbic acid. Berries have a natural sweetness and tartness, particularly fresh organic berries. If you like the mixture as is, continue below with the recipe. If it needs more sweetness or tartness, add the xylitol and/or the ascorbic acid until you arrive at the perfect taste for your palate (and for the particular berries you have, as they vary in taste).
2. Pour the mixture into an ice cube tray (silicone is best) and leave it in the freezer for 3 hours (or use an ice cream maker, if you have one). Once frozen, put the cubes back in the blender and blend briefly, just until smooth. Scoop the gelato into four bowls and enjoy.
3. *Note:* Ascorbic acid is vitamin C. You can just break open a couple of capsules if you take it as a supplement.

RASPBERRY CHOCOLATE PUDDING

SERVES 2 TO 4

Is this an unbelievably rich food that feels indulgent? Without a doubt. Is this also a nutrient powerhouse that sharpens your cognitive performance by stoking the mighty mitochondria? Hell, yes.

INGREDIENTS

4 cups full-fat coconut milk, BPA-free and guar gum–free

Up to 4 tablespoons xylitol or stevia

2 tablespoons Bulletproof CollaGelatin or 1 tablespoon gelatin

2 teaspoons vanilla extract

¾ cup Bulletproof Chocolate Powder

4 tablespoons unsalted grass-fed butter

1 tablespoon Brain Octane Oil

¼ cup pistachio nuts (optional)

½ cup organic raspberries

Shaved 85% organic dark chocolate

1. In a small saucepan over medium heat, heat 1 cup of the coconut milk, the xylitol, and the gelatin until the gelatin is dissolved. Place the remaining 3 cups of coconut milk in a blender with the vanilla, chocolate powder, butter, and oil. Blend thoroughly. Add the hot coconut milk/gelatin mixture to the blender and pulse until mixed, with or without the pistachio nuts. Pour out the mixture into a large bowl and let it set for 1 hour.
2. Serve the pudding in small coffee cups, covering the top of each dessert with a halo of raspberries. Then use a carrot peeler to shave the dark chocolate in generous curls over the raspberries before serving.

SLEEPY SNACKS

To really supercharge your sleep, try having one of these snacks before bed to give your cells the energy they need to stay powered all night long. If you wake up between three and five a.m. and can't go back to sleep, it's often because your blood sugar is dipping during the night, causing a cortisol spike, which wakes you up. This happens when your mitochondria don't get enough energy. Remember, you need lots of energy to get good-quality sleep!

HEAD STRONG TEA

SERVES 1

Herbal teas are good at soothing your nervous system so you can sleep, and adding energy with Brain Octane Oil and honey will keep your mitochondria humming all night long while you doze peacefully.

INGREDIENTS

1 cup chamomile tea or other herbal tea, such as mint

1 tablespoon Brain Octane Oil

1 tablespoon raw honey

Cool the tea, then blend in the oil and honey and enjoy. If you're not a tea person, try simply mixing the oil and honey together with collagen for a quick nightcap.

SLEEP SOUNDLY BITES

SERVES 1

These are filling and especially helpful if you suffer from energy brownouts during the night as described above.

INGREDIENTS

1 tablespoon almond, cashew, or pistachio butter

1 tablespoon raw honey

Sea salt

Mix the nut butter and honey together and form into little bites before sprinkling them with salt. These are also good any time of day for a quick energy-boosting snack, but remember that this much honey will take you out of full ketosis.

After the two-week program is over, feel free to branch out, creatively using these ingredients to create a variety of meals that are as simple or complex as you like. The most important thing is to get plenty of healthy fats, polyphenols, and neurotransmitter precursors so that your brain is running efficiently and cleanly, giving you the energy you need to be the best version of you.

11

HEAD STRONG LIFESTYLE

Eating to fuel your brain is just the beginning. Making simple tweaks to your light exposure, cold exposure, sleep, meditation, and exercise habits to support your mitochondria instead of slowly sapping your energy can have a real impact on your performance from day one. These may be small changes, but because they are things you do often, their effects really add up. After implementing these changes to your lifestyle for two weeks, you'll see clear and positive results.

LET THERE BE LIGHT

The changes you make to your exposure to light will have a profound impact on your daily performance. In order to maximize your energy and the quality of your sleep, you should avoid junk light and make sure you're exposed to the right light frequencies at the right times of day. This is incredibly important. Before you get started on the Head Strong program, take some time to complete the following steps to protect yourself from junk light.

Make the following changes to your home/office environments first.

BLOCK LEDS

You only have to do this once! Go through your house and cover all of the blue, white, and green LEDs you can find (red are okay). These might be on air-conditioning units, televisions, USB chargers, and other electronic devices. You can cover the lights with old-fashioned electrical tape, which completely blocks the LED but doesn't look so nice. Or you can use transparent TrueDark™ dots that are designed for this exact purpose. They block the harmful frequencies but still let you see the LED status. Biohacked (biohacked.com) makes these pre-cut dots, which look much more natural and unobtrusive than black tape. Plus, you can still see whether the LED is turned on (it's just a lot dimmer and a different color). I'm a backer of biohacked .com and advise them on biohacking technologies. They ship a box of curated biohacking gadgets (such as the TrueDark™ dots) that are hand-selected by me to thousands of subscribers every quarter.

MAXIMIZE YOUR TECHNOLOGY

You spend a lot of time staring at screens, and it's important to avoid taxing your mitochondria while you do so. Hack your computer light first by installing f.lux software from getflux.com. I've used it for more than twelve years. The software is free, but consider making a donation to support the cause. The software will automatically lower your screen's blue light output at night. Then go into the settings and tell the software to remove some of the blue during the day, and as much as possible at night. As you reduce your exposure to blue light, you'll see improvements to your sleep and a reduction in end-of-day brain fog as well.

Next, turn to your phones and tablets. Install the same f.lux software on your Android devices, which will naturally reduce the amount of blue light your devices emit in the evening. You can create the same effect on your iPhone, but it's a little trickier, so I added a video to help you at bulletproof.com/headstrong. This will reconfigure your iPhone so that the light from the screen won't harm your mitochondria, even during the day. Set your smartphone to the maximum warm setting without much backlighting and keep it that way all day unless you need to see precise color for your job.

Even if you use f.lux or tweak your Apple settings, it's worth your time to install blue light–blocking screen protectors on your smart-

phone, laptop, tablet, and any other devices you use. Software that dims your screen is helpful, but it can't change the LED light itself, which still gives off too much blue. The filter is there to catch some of that remaining blue spectrum so you can sleep better. You install it once and never have to think about it again. If your computer has a green or blue LED indicator when the camera is on, put a TrueDark™ dot on it so that you can still tell if the camera is on but you aren't constantly staring at a bright LED as you use your computer.

Manage your TV, too. Go into your television's settings and dim the monitor's brightness and reduce the blue tone. But because you want more brightness during the day and less at night, you may have to tweak it regularly. The easier (but not that cheap) solution is to get a fantastic HDMI box from Drift TV that permanently plugs into the HDMI input on your TV. It works by slowly removing blue spectrum light from your TV screen about an hour before your bedtime using an on-screen menu, and it restores normal color after your wake-up time.

DIM THE LIGHTS

Harvard sleep researcher Steven Lockley found that even a small amount of light at night—far less than a normal reading light produces—could reduce melatonin to change your sleep for the worse.[1] A lack of melatonin reduces your mitochondrial performance.  Install dimmer switches in as many places in your home as you can, especially in your bedroom, living room, and any other areas where you spend time before bed. That can get expensive at about $10 each, so it's okay to start with just an in-line lamp dimmer switch added to a few lamps in your house. Those are more affordable, you can buy them online, and you don't have to know how to remove a switch panel to use them.

Dim your lights two hours before bed or turn off most of the lights in your house, especially any white LEDs and CFLs. It doesn't take much LED light to confuse your body. White LEDs draw five times more insects than other outdoor lights. It confuses the bugs, and it is doing the same thing to your brain. Dimmer lights will tell your body to start producing melatonin so you can wind down and go to sleep. Bonus points if you use a few candles instead of electric lights, for their ambience and healthy light spectrum.

SET UP A SLEEP CAVE

It is essential to make your bedroom as dark as possible. Block all of the light sources you can, whether it's with blackout curtains or just pinned-up fabric. If you live in a city, put up something that doesn't allow in any outside light pollution. Seriously, for this two-week period, your bedroom curtains can be ugly. And after you see how well you sleep during this two-week period, you'll want to get real blackout curtains. You'll also understand why I sometimes tack a blanket over the windows in hotel rooms that have really bad curtains!

Next, kill your alarm clock. If you can see it, it's too bright. And it probably makes electromagnetic fields (EMF), which also impacts your sleep. Move it away from your bed and cover it, or just get rid of it. You can put your phone on airplane mode and use that as an alarm clock instead. You can also use a smartphone to track your sleep if you want; there are dozens of apps that use your phone's microphone to know when you're in deep sleep, and I keep an updated list of the best ones on the Bulletproof website.

Also upgrade your bedroom and bathroom lighting. Replace all of the compact fluorescent and white LED lights in your bedroom with lower-wattage halogen bulbs, or go all-in with a lamp that has an amber or red bulb. The amber or red bulb may look funny, but it won't disrupt your sleep like fluorescents or LEDs.

For the two-week program, try using candles in the evening instead of electric lights. This is easier and cheaper than buying new bulbs and dimmer switches. Plus, candles are purely analog and strangely relaxing.

If you use a night light for trips to the bathroom, make sure it's a red or amber one. I keep an amber night light on in the bathroom for the kids—it's the same one we carry in the Bulletproof store. Whatever you do, make sure you have no CFL (curly) or white LED bulbs in your sleeping area for these two weeks. If you have any LED indicator lights in your bedroom, even the tiny ones on power supplies, you should cover them with black tape or the TrueDark™ dots. You only have to do it once, and it will improve your sleep every night from now on.

Next, take some time to protect yourself from junk light in other environments. You're not going to be able to avoid all junk light unless you become a shut-in, and avoiding it completely isn't even necessary. Your goal is to reduce your exposure as much as you can while still living your life, and also to reduce your exposure at the times of day when junk light does the most damage. Spending the day under crappy LED lighting won't kill you, but you certainly won't sleep as well, you'll be tired, and you may crave more sugar.

And this makes an important impact on your body. A Harvard study found that people with broken circadian rhythms had higher blood sugar and lower levels of leptin, the hormone that helps you feel full. Luckily, there are simple steps you can take to protect your eyes and your mitochondria from the junk light in your office, local coffee shop, grocery store, and especially airplanes, which have some of the worst junk light around.

LOOK (AND FEEL) LIKE A ROCK STAR

Rock stars get away with wearing tinted glasses indoors—and so can you! You may get a few funny looks at first, but if you work in a bright office environment or a big-box store all day, you owe it to yourself to try it. You can tell curious friends about a study that found that exposure to six and a half hours of blue light during the day (what you get in most offices) suppressed melatonin for three hours—which is twice as much as with green light.[2] This is especially important when you're in an environment that has junk light within two hours of bedtime.

You have a choice about what kind of glasses to wear to complete your rock star look. There are cheap amber safety glasses available online that help with blue light, but they don't cover all of the spectrums that affect your sleep rhythms. The gold standard—and what I use—is a type of patented glasses with a lens filter that blocks every single one of the light frequencies documented to impact sleep rhythms. The glasses are called TrueDark™ and are from biohacked.com, the same company that makes TrueDark™ dots to cover junk light LEDs. You can still see when you're wearing these glasses, but your biology believes you're sitting in the dark. I wrote more than 90 percent of this book at night when my kids were asleep, and I was wearing TrueDark glasses the entire time.

Glasses that help your brain think it's dark will help your brain make more melatonin, which will improve your sleep. When you get more quality sleep, you'll increase your rate of neurogenesis, and you'll reduce eye damage from blue light. It's worth looking a little goofy to grow a bigger brain and wake up with more energy.

If you already know that you have light sensitivity or if you have dyslexia or other reading problems, you might consider getting the Irlen lenses I mentioned earlier. These are glasses with a custom tint specifically for your brain. You can get them from a specialist who measures your eyes to see which specific filters will give you more brain energy. I use Irlen lenses during the day, when I'm exposed to junk light, and switch to TrueDark lenses in the evening. Sometimes I wear TrueDark lenses during the day for a half hour to give my brain a break when I'm in really crappy indoor lighting, am on a plane, or just want my brain to chill out.

Wearing the right glasses indoors may or may not make you look like a rock star—or just a total dork. But it can make you *feel* like a rock star, and it will make your mitochondria perform like rock stars. And that's what really matters.

PROTECT YOUR SKIN FROM JUNK LIGHT, NOT SUNLIGHT!

Your skin is photosensitive and absorbs junk light just like your eyes do. Whenever I travel on airplanes, which have truly unnatural lighting, I wear a long-sleeve shirt, pants, and a baseball cap with a brim to protect as much of my skin as possible from the plane's junk light. When I do this, I notice a huge difference in the amount of jet lag I experience after I land.

The light you want on your skin is natural sunlight, unfiltered by windows or sunscreen, so wear shorts and short sleeves if you'll be outside. But when you're going to be locked in an office with bad lighting, go for long sleeves to give your skin and your mitochondria a break. You don't need to be perfect on this—just keep it in mind and lean in this direction. Soaking up some rays is cool when they come from the sun. It's not so cool when the rays are from artificial lighting that disrupts your mitochondria.

BALANCE WITH HEALTHY LIGHT

Morning: To compensate for the junk light you will inevitably be exposed to during the day, make sure you get some healthy light. Your best bet is actual sunshine. Go outside for at least a few minutes in the morning without sunglasses. Show a little skin to help create vitamin D_3 sulfate. Sunlight has all of the normal spectrums of light that your body expects, from infrared up to ultraviolet, and your mitochondria will work better when they get signals from this light at the right time of day. This obviously works best if you live in a warmer, sunnier climate.

What should you do if you live somewhere that isn't so tropical? I have lived in Canada for almost seven years. I get lots of sun in the summer, but it's rainy for most of the winter. Here's what I do, and what I recommend you do if you live far away from the equator: In the morning after having my coffee, I stand for ten minutes in front of a narrow-spectrum UVB tanning lamp hanging in my bathroom. Ultraviolet B radiation is responsible for creating and activating vita-

min D in the skin, and it does not damage the skin the same way that ultraviolet A radiation does.

This practice is controversial for reasons I frankly don't understand. I believe that if everyone far north of the equator did this, it would transform health care costs. It's like we have fallen into a trance. Excessive ultraviolet A radiation causes sunburns, which are associated with cancer, so we decided with no evidence that we should remove *all* sources of ultraviolet radiation. But our bodies were not designed to be in an environment without any UV light.

Exposing my skin to ultraviolet B light made a difference in how I felt, but what made the biggest impact in midwinter was actually removing the eye shields for one minute out of the ten and exposing my eyes to a small amount of UVB light from the low-pressure bulbs in the lamp (please read the warning that follows). Eye doctors and cataract surgeons will probably be aghast at that admission, because we have been trained to believe that ultraviolet radiation is bad for the eyes. It *is* bad for your eyes when you get too much of it, but it is also bad for your eyes to get none at all.

I am including this information here because it makes a giant difference in how my brain works in the winter, and it substantially reduces my visual sensitivity to junk light. There are studies showing that ultraviolet light in the eyes is correlated with higher dopamine levels in the brain. It is my hope that this area of research will receive more attention as a result of this book, but you should not expose your eyes to ultraviolet B radiation without significant research and support. You can purchase reptile lights or search "narrow-spectrum UVB bulb" to find this type of light online. Please be responsible, though—you can go blind, burn your skin, and even develop cancer if you misuse a high-powered lamp (that's why mine are all low-powered, and even those can hurt you without proper precautions!).

Midday: It's best to go outside at noon without sunglasses on for a few minutes, but if that's not possible, getting more red or violet light indoors at midday can be a helpful alternative. One of the simplest things you can do if you work indoors under bright fluorescent or LED lighting is to add some red to your environment. You are getting overdosed on blue light, so changing the ratio of blue to red can help your eyes and your brain, and your mitochondria will thank you. Sim-

ply install a red light somewhere in your field of vision. I use red LED "tape lights" on the ceiling above my desk and leave them on all day to balance out the blue in my monitor.

Nighttime: At night, red is your magic color. Minimize sources of blue and white light and use red or amber LED bulbs whenever possible, or use TrueDark glasses. I use the glasses in hotels and other foreign environments and red LEDs at home. I also use the glasses at home if I'm looking at bright screens.

You Are Electric and Magnetic

Your mitochondria are semiconductive, which simply means that they conduct electricity at varying rates. As you have read, the process of creating energy in your mitochondria is an electrical process. This means that magnets and electromagnetic fields (EMFs) affect your biology and your mitochondria.

This isn't really news. Back in 1962, Robert O. Becker observed that nerves, collagen (connective tissue), and bone were all semiconductive.[3] Then in 1984, the head of the famous Karolinska medical school in Stockholm, Sweden, published an epic $1,200 textbook on the subject and was promptly fired for disagreeing with a chemical view of our biology.

Even with the profound breakthroughs we've had in understanding mitochondria over the last few years, the exact relationship between EMFs and mitochondria is still being researched. But so far, we have abundant evidence that EMFs impact mitochondria. For instance, did you know that if you are a man and you keep your cell phone near your genital area, it will reduce sperm quality (including motility and viability) and your testosterone levels? Cell phones produce EMFs. In one study, sperm that were exposed to EMFs had fewer antioxidants and an 85 percent increase in free radicals.[4]

That is really scary, particularly because of what we know about the relationship among antioxidants, free radicals, and your cells' energy production. Sperm are some of the highest energy consumers in the body—their little mitochondria are pumping at full power to provide energy for their onetime race to an egg! If EMFs from your phone are making your sperm unviable, what are they doing to the rest of your mitochondria?

EMFs can also affect your myelin, the vital insulating lining of your nerves. In one study, rats that were exposed to EMFs had significant lesions in their myelin sheaths, a greater risk of developing MS, and neurological problems.[5] Exposure to electromagnetic frequencies also increases blood sugar.[6] How does that happen? Well, you know from everything you've read in this book that a decrease in mitochondrial efficiency results in higher blood sugar—because when your mitochondria aren't working well, the sugar in your blood isn't turned into energy as efficiently. And where is that extra sugar going to go if it's not used by your mitochondria because they've been damaged by EMF? It hangs around in your blood.

Protect yourself from EMFs by keeping your cell phone out of your pants pockets and using a headset when talking on the phone. This is another reason to turn off your Wi-Fi at night or sleep with your phone in another room. And if you want to take it up a notch, you can even install electrical filters in your house like I do. Electrical filters plug into your outlets and reduce the chaotic reverberations in your household wiring, so you are exposed to fewer highly varying EMF fields.

EXERCISE YOUR WAY TO A BETTER BRAIN

To maximize the brain benefits of exercise with the least time and effort, there are three components to focus on with your exercise plan over the next two weeks: meaningful movement, high-intensity interval training (HIIT), and resistance training.

MEANINGFUL MOVEMENT

Movement can be yoga, a walk, a hike, a bike ride, or a dance party. Just get moving, but not superfast. You can do the same thing every day or alter your movement each day. Keep this movement to a moderate intensity, and make sure you can carry on a conversation while you move. If possible, do your movement outdoors so that you stack up your benefits. You'll get healthy light while exercising, so you can grow new neurons and mitochondria and energize them all at once.

If you have access to a pool, swimming is a particularly beneficial form of exercise. Just being in water up to your shoulders increases blood flow to the brain by 14 percent,[7] and when you swim underwater the added water pressure increases it even more. Plus, holding your breath underwater is one easy form of intermittent hypoxia training that grows mitochondria. If you're really brave, combine all of these benefits with cold thermogenesis by swimming in water that's 60°F or lower.

If you choose a form of movement like yoga, Pilates, or tai chi, you get the mitochondrial benefits of basic movement and the brain-building benefits of moving your limbs across your midline. This raises BDNF and improves cross-brain communication, which walking doesn't.

Using a whole-body vibration plate, such as the Bulletproof Vibe, can do the same thing as twenty to forty minutes of exercise and in less time. You can just stand on it (or do a few basic yoga poses while on it) for ten minutes and let the vibration stimulate your mitochondria.

Recommendation: Three to five times a week, move your body meaningfully but not super-strenuously, for twenty to forty minutes. Bonus points for doing it outdoors in the morning without sunglasses to get a full spectrum of natural light at the same time.

HIIT-BACK EXERCISES

Just once each week, go outside and run four hundred yards like a tiger is chasing you. Run as fast as you can—as if your life depends on it. Then do something that feels lazy—either sit down on a bench or, better yet, lie on your back for ninety seconds. This way, your nervous system (and mitochondria) have more time to completely recover (walking around during your recovery means your nervous system recovers less). Then repeat one more time. Two four-hundred-yard sprints (followed by ninety seconds of rest each time) once a week will give you a huge boost in BDNF. How easy is that? No excuses!

Recommendation: Once a week, sprint like crazy for four hundred yards, then lie on your back for ninety seconds. Do it twice. (You can use a treadmill or stationary bike if you have to.) Don't forget to pick the pine needles out of your hair when you're done!

RESISTANCE TRAINING

In addition to your meaningful movement and your weekly HIIT-back, plan a once-weekly resistance-training workout to raise BDNF and stress your mitochondria. On this day, it's better not to use antioxidant supplements, as you want to create some stress in your mitochondria so that they will strengthen their systems. You can do resistance training directly after your HIIT-back to save time, or you can space it out by several days to make the most of your recovery. These will not be long sessions, so you have no excuse not to make time for this once a week.

This workout was inspired by Dr. Doug McGuff's *Body by Science*. Using the guidelines below, complete the following five compound movements for each workout session. Unless you're experienced with free weights, I recommend using weight machines because they are safer than free weights when it comes to reaching muscular failure. If you are not familiar with these movements, visit bulletproof.com/headstrong for a video that demonstrates safe and proper form.

1. Seated Row
2. Chest Press
3. Pull Down
4. Overhead Press
5. Leg Press

Perform only one set of each of the five movements per workout, taking each set to the point of positive muscular failure. This is the point where the weight won't move anymore no matter how much effort you apply, signifying that your muscle is completely fatigued. As you complete the movements, move the weight at a slow tempo for the duration of each set. It should take six to ten seconds to raise the weight each time and another six to ten seconds to lower it. Do not pause at the top or bottom of the motion.

The weight you use should be heavy enough that you reach muscular failure somewhere between one and a half and two minutes. If you can keep going after two minutes, add more weight. If you can't go past one minute, remove some weight. *Do not* rest at the end of each repetition. The idea is to never let the muscle relax, not even for a second.

Move on to the next movement as soon as possible without much of a break in between. The time between movements should not exceed two minutes. If you do it right, each full workout of all movements should take about twenty minutes. That's only twenty minutes a week!

HACK YOUR SLEEP

The junk light hacks and nutrition plan will help you sleep better than you probably have in years, but there are a few more things you should do to make sure you get the best-possible quality sleep. These simple steps will help your mitochondria clean house during the night so you can wake up fresh and ready to kick ass all day long.

- **Switch to Decaf Coffee After Two P.M.** As much as I love coffee, I don't like how it affects my sleep when I drink it after two p.m. The caffeine can keep many people from sleeping soundly when they drink it late in the day, so cut out the caffeinated French press or cappuccino after dinner for the next two weeks. Decaf is fine—just watch out for mitochondrial toxins in untested decaf. Even countries with mold standards for coffee allow twice as much mold toxin in decaf!

- **Don't Work Out Before Bed.** Do not do your resistance workout or HIIT-back within two hours of going to sleep because it will energize you and keep you awake. You can, however, do your meaningful movement before bed. In fact, some relaxing yoga right before going to sleep may help you wind down.

- **Go to Sleep Based on Your Chronotype.** It turns out that your genes influence your circadian rhythm in a meaningful way. About 15 percent of us naturally stay up late, about 15 percent of us naturally wake up early, about 15 percent of us don't sleep well at all, and the other 55 percent follow the normal daily cycle. If you are one of the 55 percent, go to bed by eleven p.m. to avoid the cortisol surge that can come around that time and keep you from sleeping. This is the famous second wind that you may have taken advantage of in college when cramming for finals. It may have helped you then, but it will harm your sleep quality now. If you are an early riser or a night owl, you can shift that time by an hour in either direction.

  The fact that your mitochondria follow your circadian rhythm really matters. This is new research based on the work of my friend Dr. Michael Breus, the sleep specialist from *The Dr. Oz Show*. If you're not sure of your sleep type, you can take  a free quiz to determine your chronotype at bulletproof.com/chronotype or read *The Power of When* by Dr. Breus, which I highly recommend.

- **Hack Your Sleep with Honey.** Have up to 1 tablespoon of raw honey before bed on an empty stomach. Your brain uses liver glycogen (carb storage) at night, and raw honey replenishes this supply better than other carbs, so you can create stable glucose levels for hours. Stable glucose means happy mitochondria, as they are busy repairing at night. Many people take the honey with Brain Octane Oil so that your brain can burn glucose (from the honey) and fat (the ketones from the Brain Octane Oil) while you sleep for maximum efficiency.

- **Go into Airplane Mode.** When you go to sleep, turn off the Wi-Fi on your phone and put your phone in airplane mode (or leave it in another room while you sleep). This will protect you from EMFs and any lights or sounds that your phone might make during the night. If you're worried that people won't be able to reach you all night, you need to relax. I am officially giving you permission to turn your phone off when you go to sleep. It will make you stronger the next day so that you can better face whatever emergencies you slept through during the night. I'd suggest turning off your Wi-Fi router at night as well—you'll be amazed at the improvements to your sleep. I use a vacation timer that automatically cuts power to my WiFi router at night.

- **Just Breathe.** Performing a simple breathing exercise before bed will lower your cortisol levels and turn off your fight-or-flight response to help you sleep, and your mitochondria can detect this difference. There are two simple breathing exercises that I find most helpful when going to sleep—in fact, I don't know how to stay awake after doing these. They are called a box breath and an ujjayi breath. You can do either or both each night during the next two weeks.

  To do a box breath, sit down in a comfortable chair with your feet on the floor and your hands in your lap, or do it in bed lying flat on your back. Close your eyes, close your mouth, and slowly breathe in through your nose as you count to four (or more). Hold your breath for another count of four (or the same number as your inhale), and then exhale through your mouth for an equal final count. Finish up by holding your lungs empty for the same count. Then repeat several times. You can watch a video tutorial at bulletproof.com/headstrong.

  An ujjayi breath is used in yoga, tai chi, and Taoism and is sometimes called an "ocean breath" because of the sound it makes. It's hard to describe it in writing, but I'll do my best—and you can visit bulletproof.com/headstrong to watch a video of a properly done ujjayi breath. Sitting comfortably, or lying down, breathe slowly and deeply through your nose, filling your abdomen first. Tighten your throat as you inhale, almost like you would if you were going to snore or snort. It should sound like the ocean in your ears. When your lungs are full, breathe out through your nose the same way, with a tightening of your sinus area to make the sound, as if you're trying to fog up a mirror.

Do either one of these breathing techniques for four minutes before bed, and you'll sleep the relaxed sleep of a very tired baby.

MITOCHONDRIAL MEDITATION

I have been meditating for years, and in many different ways (including with lots of sensors glued to my head), but I wanted the Head Strong meditation to specifically benefit your mitochondria. So I sought out a five-thousand-year-old Chinese energetic medicine practice used to protect China's emperors. It translates to New Life Energy, and Dr. Barry Morguelan is one of twelve remaining grandmasters of this form of training in the world, as well as a highly respected UCLA surgeon. He spent years at the top of a mountain in China while he was growing his Western medicine practice, studying and doing the kind of training where you sit with your shirt off and learn to melt snow around you with your body heat. Talk about mitochondrial function!

Barry is one of the most powerful energy medicine workers I have ever met; he has helped me in many tangible ways, he has treated the presidents of multiple countries, and he flies in for Tony Robbins's big events to help keep Tony's impressive energy at its peak. I could not imagine a more qualified person to design this meditation for you.

The entire practice of New Life Energy is built around increasing and controlling the mitochondria, allowing them to create more energy and to direct it with more power. This is a core meditation from the practice, one Dr. Barry selected specifically for mitochondrial function. There are no clinical studies of this specific form, but I'm happy to go with five thousand years of observation from the creators of this school of energy medicine. And if I'm wrong, it's still an amazing meditation!

A meditation like this is best experienced with headphones, and there's a recording with Dr. Barry available for free at bulletproof .com/headstrong. Here is the written version, but you owe it to yourself to try the audio version.

Find a comfortable chair that supports your back or lie on your back. Close your eyes and uncross your arms and legs. Relax your hands on your thighs. Take a deep breath and breathe in the whole universe. Scientists have noticed that this is a fluid universe and that it communicates with us in many, many ways that are still undiscovered and, yes, as you breathe, there is motion everywhere. Nothing is standing still. So enjoy the excitement of taking a deep breath all the way down to your diaphragm.

Now take a deep breath and let it go all the way down to your feet. Now take a breath and notice that you can direct the breath like a beam that you can send all the way down through your diaphragm, through your pelvis, through your legs, through your feet, and into the ground. Then, enjoy noticing that as you breathe all the way down into the earth you can relax, and the beam of the breath can make a pathway back up from the earth through your torso up to your lungs. Now, notice that you can take an even deeper breath and follow that pathway again all the way down, down, down into your abdomen, into your legs, into the ground, and burrow down deeper into the earth through sediments, the shale and rocks, and into oil and water collections.

Notice you can follow your breath in a pathway that continues to keep tunneling downward further and further. Send that beam all the way down and notice that it doesn't take that long for it to work its way through deeper and deeper into the earth. As you breathe again, follow the breath as it continues to gain momentum and strength and watch it create a beam that goes down through your body, further and lower and deeper into the earth, this time about a mile down. Then watch it turn around and come back up all the way, retracing its pathway.

As it comes up, the pathway is already open for the beam to return much faster. It comes up faster and faster and faster, up into your chest, up into your head, and can actually come up even a foot or so above your head. Experience the beam with your eyes closed as it explodes above your head and then notice that it really has some power now. Breathe again, deeply, and send the beam down again—all the way down through your body, through your legs.

Keep following the momentum and the power behind it now. You can follow your breath like a beam as it passes down further and further, nearly to the center of the earth. And, then, once it reaches that point it lightly touches the center and then circles around and starts to come right back up much quicker. The beam comes up faster and faster, farther and farther until it passes up through the earth and into your feet, into your body, into your head, and then goes about three or four feet above your head. You may notice that as you keep breathing the beam gets stronger and appears thicker, often more golden and platinum colored.

As it becomes easier to breathe and direct this beam, you can direct it to retrace its pathway down through your body, through the earth and into the center and then through many layers until it finally breaks through the crust of the earth on the other side of the planet, creating a showering effect. Then as you continue to breathe, you notice just as easily, the beam returns from the other side of the earth's crust all the way back through, and up through your body and your head. Sit there for a while and relax and just breathe at a comfortable pace as you watch the beams going down from you making their path through the earth out the other side. Then watch them turn around and come all the way back to you like a column. This column made by the beam appears more golden and platinum colored as it makes its way back up to you.

On one of these times that it comes back to you, you can experience being lifted up higher and higher into an incredible blue sky. As you start to go higher and higher you recognize your ability to climb a pathway up a mountain, and there seems to be less resistance. You can move even faster now as you go higher and further on the path. It doesn't take long until you encounter an old hut along this well-worn path that stands very beautiful in the sunlight.

Notice there are more and more attractive green fields with farmers working, and the fragrance of the fields becomes more noticeable. As you look just off to the side of the pathway, you see the platinum gold beam that's been supporting you in increasing amounts on your adventure. The beam almost appears like a sleigh, one that you can sit on and ride as it goes down and around the mountain. Above you is a beautiful branch from a very large tree that has these very translucent, yellow leaves that float back and forth in the wind. As one of these leaves slowly makes its way down to this grassy knoll next to you, it's a great time to rub your hands together eight times, and then put your palms over your eyes and just relax, taking it all in. It's not necessary to feel anything. There can be a sensation of warmth, or coolness, or prickliness, or no sensation whatsoever. It is all there for you.

Then, after about a minute, rub your hands together again about eight times to get some friction between them, and then put both hands over the center of your chest, palms next to each other, or on top of each other. Relax and take that in now as you breathe calmly. After a minute of enjoying the chest position, once again rub your hands together eight times and gain some friction, and then put both hands over your belly button. Your hands can be side by side or on top of each other.

As you take in all that's there, let it pass through you. You may feel something different each time, or you may feel nothing at all. Every time will be different! After resting in this abdomen position for a minute, release a big smile, take a deep breath, slowly open your eyes, and stretch. From there, enjoy taking on your next opportunity for success.

Recommendation: Practice this meditation once a day throughout the two-week program, ideally using the audio version at bulletproof .com/headstrong for maximum effects.

COOL OFF

For the next two weeks, you are going to benefit from the cold. Every morning at the end of your shower, turn the water all the way to the coldest setting for the last thirty seconds. Stand there and let the cold water hit your body, especially your face and chest, which are two big activating areas for cold thermogenesis. Yeah, you're not going to like this, especially during the winter. Sorry. But it's only for two weeks, and you *will* like the amazing burst of energy you get as a result.

If you don't shower every day (hey, it's none of my business) or if you want to take it up a notch, try sticking your face into a bowl or sink full of ice water for as long as you can stand it. At first, it may only be for ten seconds. That's okay—you can work your way up. By the end of two weeks, you'll be able to tolerate it for as long as you can hold your breath. This allows you to have intermittent hypoxia and cold thermogenesis at the same time!

An easy way to do this is to fill a shallow pan or dish with water and put it in your freezer. When the water is frozen solid, take the pan out, add water on top of the ice, and stir it up to chill the water. Hold your breath and put your face into the pan until you can't handle the discomfort from the cold anymore.

The optimal temperature for the water is less than 50°F (because you want to get your skin temperature as close as possible to 50–55°F). You can use a thermometer to check the temperature of the water until you are better at estimating it. I ended up using a snorkel so I could keep my face submerged for up to five minutes at a time. While this is energizing, it's also very relaxing. In fact, performing this trick before bed every night will drop your body temperature and help you fall asleep faster and sleep deeper. I noticed profound changes within a week when I started experimenting with this a few years ago.

Whew! That's a lot to keep track of, so here is a quick checklist to make sure you're remembering all of your Head Strong lifestyle practices. I also created a one-page checklist with a weekly schedule for you, which is available to download for free at bulletproof.com/head strong.

One and Done

- Set up your lights
- Set up your sleep cave
- Maximize your technology
- Invest in tinted lenses

Daily

Any time of day

- Cold shower or facial ice bath
- Mitochondrial meditation
- Meaningful movement

In the Morning

- Get natural sunlight—at least ten minutes

In the Afternoon

- Get natural sunlight or violet light
- Switch to decaf after two p.m.

At night

- Take 1 tablespoon of raw honey
- Dim the lights two hours before bed or switch to candles
- Do a breathing exercise
- Turn off your Wi-Fi and router
- Put your phone in airplane mode
- Go to bed by eleven p.m.

Once a Week

- HIIT-Back Exercise (not within two hours of bed)
- Resistance Workout (not within two hours of bed)

Are you excited for the next two weeks to be the most energy-packed and productive weeks of your life? Once you see what it's like to feel this good, it's hard to go back to what used to be your normal. And you don't have to. I've been following this program for years, and I have no intention of going back to the fuzzy, forgetful brain I used to have. I hope you'll be with me and stay forever Head Strong.

12

HEAD STRONG SUPPLEMENTS

You don't have to take supplements to make your mitochondria stronger. But there are some supplements that dramatically move the needle more than is otherwise possible and others that can overcome specific obstacles that are preventing your mitochondria from making maximum power. If you already have the symptoms of mitochondrial decline, these supplements are the big guns to quickly bring you back online.

I am well known for taking dozens of supplements every day to maximize my energy and performance—but then again, I'm a professional biohacker with a stated goal of living to the age of 180. I'll tell you in 137 years whether or not I make it, but in the meantime I will do every little thing I can to get my mitochondria working to their fullest potential. Some of these supplements have absolutely changed my life by helping me find the limitless energy I sought for so long.

Based on all of my experimenting, I've put together a list of the most important supplements that can dramatically ramp up your performance. I recommend you take these throughout the two-week program and beyond. High-end racing cars require more maintenance and better fuel than your uncle's old beater. I may drive a pickup truck, but my mitochondria are high-performance machines, and I take care of them accordingly.

Note: You can order most of these supplements online or pick them up at a local supplement store. Some are highly specialized, and some I had to create over the past fifteen years to get my biology where I wanted it to be. I spend my creative energy making stuff I desperately want that I can't get anywhere else, and I'm sharing it with you here so you can decide what is right for you.

Though I recommend using all of these supplements during the two-week program, I have broken them down into three categories based on their cost, effects, and how immediately your mitochondria turn on after taking them.

LOW-IMPACT SUPPLEMENTS

You won't feel a noticeable boost the first day you use these, but they have long-term benefits that are worth their relatively low cost.

CAFFEINE

Caffeine itself (not just coffee) is an energy booster and cognitive enhancer. It may help ease cognitive decline and lower your risk of developing Alzheimer's disease by blocking inflammation in the brain.[1] According to research done by Professor Gregory Freund at the University of Illinois, "We have discovered a novel signal that activates the brain-based inflammation associated with neurodegenerative diseases, and caffeine appears to block its activity."[2] Caffeine also increases insulin sensitivity in healthy humans,[3, 4] which is extremely important to sustained energy.

A small number of people are genetically slow to process caffeine—you may be one of them if all forms of caffeine (soda, chocolate, tea, coffee) make you feel awful. If this is the case for you, it might be best to avoid caffeine entirely.

You can get caffeine in pills, of course. But I stick to getting mine the old-fashioned way—in hot black liquid.

RECOMMENDED DOSE: 1 to 5 cups daily

FORM: Coffee

TIME OF DAY: Before two p.m.

COENZYME Q_{10} (COQ_{10})

You make this antioxidant in your mitochondrial membranes. Its job is to carry electrons into the mitochondrial inner membrane, where it is oxidized to create energy and protect against oxidative stress.[5] In other words, it helps your mitochondria produce energy more efficiently while protecting against the oxidative stress that energy production causes.

RECOMMENDED DOSE: **30–100 mg daily**

TIME OF DAY: **With Bulletproof Coffee or another source of fat for maximum absorption**

BRANCHED-CHAIN AMINO ACIDS (BCAAS)

BCAA supplements contain the amino acids isoleucine, leucine, and valine. These amino acids boost mTOR, which helps control inflammation and plays a key role in promoting cell growth and preventing cell death. BCAAs also suppress cortisol, which helps fight inflammation even more. And they can plug into the Krebs cycle to boost energy.

The problem with most BCAA supplements is that BCAA powder tastes horribly bitter and doesn't mix well with water, so many manufacturers add artificial sweeteners and dissolving agents to it. Stay away from products that contain the artificial sweeteners aspartame, acesulfame potassium (ace-K), and/or sucralose. Xylitol, erythritol, and stevia are the best sweetener options, and sunflower lecithin and (organic, non-GMO) soy lecithin are the best dissolving agents. If you don't like sweeteners (or if you're a masochist), you can drink unflavored BCAAs.

RECOMMENDED DOSE: **5 g daily**

TIME OF DAY: **Before, during, or right after exercise**

VITAMIN B_{12} AND FOLINIC ACID

Why am I listing these two supplements together? They are intrinsically connected. Many people are deficient in vitamin B_{12}, which can protect against dementia, increase immune function, maintain nerves, and regenerate cells. Your brain needs B_{12} in order to thrive. Folate deficiency can also cause mental symptoms, although lack of B_{12} is more likely to be a problem. Folate and B_{12} are both required for mental function, and a deficiency in one produces a deficiency in the other. However, folate will not correct a B_{12} deficiency in the brain. In fact, if you make the mistake of treating a B_{12} deficiency with folate, you can get permanent brain damage. That's why I take them together.

So why is folinic acid listed here instead of folate? Folinic acid is a metabolically active form of folic acid that does not require an enzymatic conversion. Approximately one-third of the population does not have the genes that allow them to process folic acid. Folic acid builds up in their bloodstream and interrupts cellular metabolism. This is why I always recommend taking folinic acid, not folate.

There are three forms of B_{12}, but the most common form in supplements (called cyanocobalamin) doesn't work for many people. And while the commonly sold form of folic acid works great for two-thirds of us, it builds up to toxic levels in the rest of us. That's why I recommend the specific forms of B_{12} and folate below.

RECOMMENDED DOSE FOR B_{12}: > 5 mg methylcobalamin or hydroxycobalamin daily

RECOMMENDED DOSE FOR FOLATE: > 800 mcg 5-methyltetrahydrofolate (5-MTHF) or folinic acid daily, *not* folic acid

TIME OF DAY: Doesn't matter, but it's sublingual (situated under the tongue), so you'll have to suck on it for a while

MAGNESIUM

Your body uses magnesium in over three hundred enzymatic processes, including all of those involved in ATP production. Magnesium actually reverses the effects of stress on your brain and can therefore increase your memory and cognitive function. Low magnesium means low brain energy, which is a serious problem. Other symptoms of magnesium deficiency include heart arrhythmias, tachycardia, headaches, muscle cramps, nausea, metabolic syndrome, and migraines. (Notice how all of these are linked to mitochondrial function!) It's also associated with cardiovascular disease, diabetes, asthma, anxiety disorders, and PMS (premenstrual syndrome).

Sadly, about 68 percent of Americans are below the already low U.S. recommended daily allowance of magnesium.[6] Due to soil depletion and poor farming practices, it's almost impossible to get enough magnesium from your diet. Given that this is an affordable supplement that's so widely necessary in the body, there's no reason not to take it.

Previously, I've recommended taking magnesium at night for sleep. I still do this because it helps with relaxation. But magnesium is a circadian mineral—your levels change with the time of day. The natural peak levels of magnesium happen at noon, and a recent study from Cambridge University found that magnesium regulates cellular energy over the daily cycle and actually helps to set your circadian rhythm. In short, the more magnesium you have during the day, the more ATP, energy, and efficient sleep you have![7] I've started taking my magnesium in the morning with my coffee, followed by a little more at night for sleep.

RECOMMENDED DOSE: **600–800 mg daily**

FORMS: **Citrate, malate, glycinate, threonate, or orotate**

TIME OF DAY: **Mainly in the morning, with a little at night if needed for sleep**

VITAMIN D_3

Vitamin D_3 is a well-known vitamin, one I've recommended for a very long time because it has so many positive effects on the body. It acts on over one thousand different genes, it can actually raise testosterone slightly,[8] it helps you produce human growth hormone,[9] and it moderates immune function and inflammation. It's no coincidence that this is one of the few vitamins we can make on our own just by sitting in sunlight. When I interviewed Dr. Stephanie Seneff, a senior researcher at MIT, she said that vitamin D_3 from sunlight is superior to any other form because the UV light sulfates the vitamin D_3 in your system, thereby activating it.

In 2014, my neighbors at the British Columbia Children's Hospital in Vancouver published a study about vitamin D_3 and mitochondria, which found that "vitamin D deficiency, when corrected, can lead to an improvement in mitochondrial function. There is a link between vitamin D and mitochondria of human muscle."[10]

So get some sunlight—and take your D_3. It works even better if you pair it with vitamin K_2 and vitamin A.

RECOMMENDED DOSE: **5,000 IU daily for adults, 1,000 IU/25 pounds of body weight daily for kids, plus twenty minutes of direct sunlight on unprotected skin per day, without sunglasses, or ten minutes of UVB tanning lamp per day (note: UVB tanning lamps don't cause tanning damage the way UVA lamps do)**

FORM: **D_3**

TIME OF DAY: **Morning**

MEDIUM-IMPACT SUPPLEMENTS

The impact of these supplements will grow over time. You probably won't feel much on the first day, but it won't take long before you notice a dramatic increase in energy.

ACTIVATED COCONUT CHARCOAL

Activated charcoal is a form of carbon that has a massive surface area and a strong negative charge. It's been around for thousands of years and is still used in emergency rooms today to treat poisoning. Charcoal binds to chemicals whose molecules have a positive charge, including aflatoxin and other polar mycotoxins and common pesticides. Once the chemicals attach to the charcoal, you can pass them normally (i.e., poop them out). Charcoal can bind to the good stuff, too, so I don't recommend taking it within an hour of other supplements or medications.

Removing toxins that directly inhibit mitochondrial respiration is a good idea. If I start to feel like I've eaten something that's slowing my mitochondrial function, I always head for my charcoal, and it makes me feel better quickly. This works for my kids, too—if they have a meltdown after a meal, charcoal works wonders.

RECOMMENDED DOSE: 1 to 4 capsules daily, not with any medications

TIME OF DAY: Before bed, to bind any toxins that might impact your mitochondria

SHAMELESS PLUG: I manufacture a special form of activated charcoal called Bulletproof Upgraded Coconut Charcoal that has an ultrasmall particle size and goes through a process to remove heavy metals. It is also environmentally sustainable and comes from pure coconut shells.

WARNING: Charcoal can constipate you! Following the Head Strong food plan should help you combat that.

CREATINE

Creatine has been a workout staple for decades, but the guys in the weight room probably don't realize that creatine works by increasing ATP production in your mitochondria. This is what gives your muscles an extra boost of energy. Creatine also benefits your brain. In one study, creatine supplements caused a significant increase in working memory and intelligence.[11] In Head Strong, we're not using it to "get big"—we're using it to make your mitochondria work better.

You have a couple of options when it comes to creatine supplements. The original and best-studied form is creatine monohydrate. It's very effective—once you get it to a high enough concentration in your muscles. To do that, you have two options:

- Take 5 g daily and wait a month for it to kick in.
- Do a "loading phase" in which you take 20 g daily (four doses of 5 g) for a week, then 5 g daily to maintain after that.

On the two-week Head Strong program, I recommend the loading phase so that you can feel the creatine's effects right away. If you do this, be sure to drink a *lot* of water. Creatine increases muscle hydration, so you should take in much more water than you normally would to counteract the effect of extra water being pulled into your muscles. It's not uncommon to experience a mild dehydration headache and bloating during the loading phase. This should go away when you finish loading and drop down to a normal dose.

Creatine gets into your muscles more effectively if you pair it with a tiny amount of glucose. You don't need much, so I recommend taking it with a quarter of a teaspoon of raw honey.

RECOMMENDED DOSE: 20 g daily (5 g four times daily) for one week, then 5 g daily to maintain

TIME OF DAY: Morning (or throughout the day for the loading phase)

WARNING: If you have kidney problems, check with your doctor before taking creatine.

KRILL OIL

Packed with DHA, EPA, and astaxanthin, krill oil helps fight inflammation, maintain your brain's structure, and keep the neurons in your brain communicating freely. With better communication in your nervous system, you're going to be smarter, sharper, and stronger. Krill oil is special because it can be absorbed directly by your brain without the additional processing that normal fish oil requires. It also has approximately forty-eight times more antioxidants than fish oil.

Astaxanthin, the compound that makes salmon red, has special mitochondrial powers—it protects mitochondria from oxidative stress and maintains a high mitochondria membrane potential. In other words, it gives your mitochondria more resilience and power.[12] Bonus points if you take it with some gamma linoleic acid (GLA) from borage seed oil or evening primrose oil. In combination with krill, GLA helps strengthen cell membranes.

RECOMMENDED DOSE: 1,000 mg daily

TIME OF DAY: With Bulletproof Coffee or fatty meals for best absorption

POLYPHENOL BLEND

We recently learned that our bodies use polyphenols to create melanin, which can turn water into energy and oxygen, thereby directly benefiting mitochondria. This shocking discovery may explain why polyphenols, the building blocks of melanin, are so important. On top of that, polyphenols feed healthy gut bacteria found most often in lean people (and these are the kinds of gut bacteria you cannot buy as supplements). Additionally, polyphenols protect mitochondria from oxidative stress. As you can see, their benefits are far-reaching!

According to Dr. Barry Sears, author of the famous Zone Diet, we should be getting at least 2,000 mg of polyphenols per day. But studies show that humans normally get between 100 mg and 1,000 mg per day.[13] That's a big difference. And because different polyphenols do different things, you want to make sure you get a wide variety when upping your levels. There are several hundred types of polyphenols in edible foods and thousands more that we can't eat. As you already know, coffee is the single largest source of polyphenols in the Western diet. I recommend you drink coffee made with a metal filter, not paper, as this increases the polyphenol count and allows the anti-inflammatory coffee oils to remain in the coffee. There are countless studies online documenting the benefits of coffee (even decaf), and I believe polyphenols and melanin are the primary uniting reason behind those findings.

In studies, the benefits of coffee are highest at five cups per day, regardless of whether it's decaf or not. Depending on the brewing method, the polyphenol content per serving is 200–550 mg. Green tea has polyphenols, too, but only 100–120 mg per serving, and they're different polyphenols.[14] Chocolate has less than that, and red wine has far less than chocolate. So even if you do what I do—drink one or two cups of caffeinated coffee and three cups of decaf per day—you might get enough polyphenols, but you'd still be missing out on variety.

After a lot of research and carefully creating a Head Strong diet that's rich in dark blue, red, purple, and green foods, I came to the conclusion that I'm unlikely to get the most effective dose of broad-spectrum polyphenols from diet alone. So on top of a vegetable-heavy, high-fat diet, I supplement with a broad-spectrum mix of polyphenols from tart cherry anthocyanins (to protect neurons),[15] pomegranate polyphenols, grape skin polyphenols, resveratrol, blueberry polyphenols, bilberry, green tea extract, and pterostilbene. I also take 200 mg of coffee fruit (*not* green coffee bean) extract, which contains a complex mix of polyphenols clinically shown to increase BDNF levels in blood. This extract is the base of NeuroMaster™—the supplement to support neurogenesis and neuroplasticity.

RECOMMENDED DOSE: Here's how to get at least 1 gram, preferably 2 grams, of polyphenols per day:

- **Drink 2 to 5 cups of coffee (mold-free to avoid inhibiting mitochondria) and/or 2 to 6 cups of green tea**
- **Eat 1 to 2 servings of dark chocolate**
- **Consider 100–300 mg green tea extract and 100 mg resveratrol supplements or the broader-spectrum mixed polyphenol supplements mentioned above. There is also a Head Strong formula available that combines the above types of polyphenols.**

TIME OF DAY: Anytime

SPROUT EXTRACT

It turns out that some plants make very useful chemicals when they are sprouting, and we can take advantage of that in supplement form. For example, broccoli sprouts make a chemical called sulforaphane, which activates liver detox enzymes and powerfully protects your inner mitochondrial membrane.[16] I take this one every day.

There's just one problem—the sulforaphane that's available online requires activation by gut bacteria that are not common, so you probably won't see any benefit from it. You can either buy an enzyme-activated sulforaphane or make sure you eat a bite of raw radish or cruciferous vegetable (like a broccoli stalk) in order to have the right living enzymes in your gut to activate the supplement. I always eat a bite of my broccoli raw, and we use radish as a garnish at our café, called Bulletproof Coffee, for this reason.

RECOMMENDED DOSE: **10 mg sulforaphane daily with raw cruciferous vegetable or raw broccoli sprouts, or an enzyme-activated supplement**

TIME OF DAY: **Anytime, on an empty stomach**

HIGH-IMPACT SUPPLEMENTS

These are the most powerful supplements for your mitochondria. Use them when you need maximum energy to feel your full power.

KETOPRIME™

In chapter 2, you learned about how your Krebs cycle accepts glucose or ketones from fat, strips off their electrons to make ATP, and then sets up the process to start again. If there is a problem with *any* of the steps in your Krebs cycle, the final step will fall short and you'll have less energy. The molecule you need for the final step is called oxaloacetic acid (OAA)—and when you have extra, you can make ATP more easily. This unique molecule is actually a type of ketone, one that must be present in order for your body to use the ketones you get from a zero-carb diet or from using Brain Octane Oil.

The problem with OAA is that it's unstable, so it wasn't usable as a supplement—until recently, that is. KetoPrime is a new, stabilized form of OAA that has a dramatic impact on your energy. It is packaged in lozenge form along with co-factors that allow the OAA to do its job in your mitochondria. The KetoPrime formula mimics the mitochondrial impact of restricting calories, which tells your genes to make you live longer.[17] It also increases the precursor NAD+ that mitochondria need to make more energy and protects the brain from excessive glutamate, actually transforming harmful excess glutamate into fuel for neurons.[18] In animal studies, OAA has shown to be more powerful than other supplements that protect your brain from glutamate.[19] Studies have also shown that it protects mitochondria from environmental toxins and free radicals.[20]

KetoPrime is one of the most powerful mitochondrial enhancers I have ever found. It "primes the pump" for the Krebs cycle, and, since it's water soluble, it can reach directly into your brain.[21] Increasing mitochondrial function can help everywhere in your body, especially before migraines or PMS even have a chance to get started.

RECOMMENDED DOSE: 100 mg sublingually, one or more times per day as needed, up to ten times a day

TIME OF DAY: Before workouts, or anytime you feel physical symptoms of mitochondrial dysfunction (headaches, tiredness, brain fog, weakness, eye stress, etc.). It helps mitochondria work at night, so it also works before bed. Just be sure to brush your teeth afterward, as it has vitamin C in it, which should not sit on your teeth all night.

GLUTATHIONE

Glutathione is your body's master antioxidant in the liver, but it has an impact on every cell in your body. In fact, mitochondrial glutathione is called a "key survival antioxidant."[22] It protects you from oxidative stress and heavy metal damage and supports the liver enzymes that break down mold toxins and heavy metals. And in the mitochondria, it stops damage from free radicals that could otherwise result in cell death or damage.

Your digestive system will destroy normal glutathione by digesting it, so you have to either take it intravenously (at the offices of a functional medicine practitioner), or you can swallow special forms that bypass your stomach digestion. Since IV (intravenous) glutathione is expensive and it involves needles, most people opt for an oral form.

The oldest and cheapest oral form is a combination of the amino acid supplement N-acetylcysteine (NAC) and vitamin C, which gives your body the ingredients to build glutathione on its own. This process is limited and inefficient, but it's the most affordable option. The second form of glutathione supplement to hit the market is called liposomal glutathione, where the glutathione is wrapped in a layer of fat that helps escort the antioxidant into your tissues. Sadly, liposomes are normally absorbed only in the top few inches of your GI (gastrointestinal) tract, so you have to hold it in your mouth for a while, and once you swallow it, it doesn't absorb well.

To hack this problem, I created Bulletproof Glutathione Force, a form of liposomal glutathione with an added bioactive molecule called lactoferrin, which allows the supplement to be absorbed into your body throughout a lot more of your GI tract. The trick is all in the efficient delivery system, and in this way your body can actually use what you swallow!

Glutathione isn't something you should take every single day—I skip one to two days a week because I don't want my body to get used to it and then reduce its own production of natural glutathione.

To be clear: IV glutathione is best, but it's inconvenient and pricey. Any of the forms can help you, and I encourage you to try one of them on the Head Strong program so that your mitochondria are more efficient and your body can more rapidly eliminate toxins. Glutathione is magic for helping with hangovers, too!

RECOMMENDED DOSE: 500 mg or more daily

TIME OF DAY: Morning or night on an empty stomach

NOTE: Skip one or two days per week to avoid building tolerance.

ACTIVEPQQ

One of the most exciting new ways to increase your mitochondrial function is to use the compound pyrroloquinoline quinone (PQQ). If you can successfully get PQQ past your digestive system, it has a measurable impact on mitochondrial function and can even cause mitochondrial biogenesis. PQQ also functions as an antioxidant, protecting against inflammation and oxidative stress. It can increase mitochondrial density to give you more energy,[23] reduce inflammation,[24] boost metabolism,[25] improve fertility,[26] and improve learning and memory ability.[27]

Four years ago I started taking 30–40 mg of the common form of PQQ every day. However, unlike other mitochondrial energizers, I never felt any effect. After ten weeks of 40 Years of Zen training to enhance my self-awareness and a lifetime of intermittent mitochondrial function because of environmental toxic mold, I can feel when my mitochondria are working well. At the high doses I was taking, PQQ was costing me a few hundred dollars a month, and I became convinced that I was wasting money and time on this booster, even though the studies on it were so promising.

It turns out that I was partially right. The likely reason I didn't feel an energy boost from megadoses of PQQ over an extended period of time is that the disodium salt form of PQQ that is commonly sold becomes inactivated in the stomach. All that expensive PQQ I was taking never even made it to my mitochondria! I set out to synthesize PQQ directly, targeting an acid-friendly form of PQQ. The result? A noticeable energy boost even at a lower dose. That is how Bulletproof ActivePQQ™ was born, and the supplement Bulletproof Unfair Advantage.

Unfair Advantage is the only product on the market today that delivers the active form of PQQ that bypasses your stomach acid. It's the most important mitochondrial supplement I've ever created, and I'm really excited about it. As soon as I started taking it, I felt a dramatic increase in my overall energy levels, especially my ability to focus and perform during times of stress. It has been a game changer for me, and I hope it has the same impact for you. Feel free to experiment first with generic PQQ to see if you get any benefits—you'll save a few dollars that way if it works for you.

One hack that's really helped me is taking this before bed. The glymphatic system that cleans your brain during sleep requires mitochondrial energy, and supporting it with PQQ helps me sleep efficiently.

RECOMMENDED DOSE: 10–40 mg daily

TIME OF DAY: During an energy slump or before sleep

Remember, all of these supplements are optional. You will notice a huge difference in the way you feel just by following the meal plan and starting the new habits in the previous chapters. But adding some or all of these supplements will undoubtedly push your performance further while working synergistically with the rest of the Head Strong program. I hope you'll try them, and I hope that they make as big a difference for you as they have for me.

13

BEYOND THE LIMITS

The two-week program in chapters 10, 11, and 12 is designed to make you Head Strong. If you like the way you feel, which I trust you will, you can stick with that program indefinitely. But if you want to take your performance to an even higher level after completing the two-week program, the hacks in this chapter are for you. These are not meant to replace the hacks in the original program. Instead, try a combination of these next-level hacks in conjunction with the Head Strong program to further enhance your results and push your performance beyond the limits.

I've done every one of these hacks as part of the research for this book. Some of them are a bit crazy. Some are expensive. And some are pretty simple. But they are all groundbreaking, and they all tie back to your mitochondrial performance.

DEEPER SLEEP HACKS

Hopefully you're already getting better-quality sleep than ever before, but there are additional tricks to reach a deep, restorative state of sleep even more quickly. Here are some of my favorites.

MAGNETIZE SLEEP

We know that mitochondria are semi-conductive, and we know that magnets affect the energy production in each cell. This knowledge helps explain how pulsed magnetic fields can have a profound effect on sleep and the brain. And the following technique, called transcranial magnetic stimulation, was designed to take advantage of this connection.

Transcranial magnetic stimulation effectively treats insomnia by stimulating the brain's production of serotonin, melatonin, and other neurotransmitters that are required for a good night's sleep. To receive the stimulation, you hold a device containing an electromagnetic coil against your forehead while short electromagnetic pulses are administered through the coil. The magnetic pulse easily passes through the skull and causes small electrical currents to stimulate nerve cells in the targeted brain region.

This technique is quickly gaining popularity. There are now clinics where you can receive the treatment, and there is already a Kickstarter campaign for a home version. I started using pulsed electromagnetic frequency devices on my head occasionally during sleep five years ago. If your brain isn't doing what you want it to do, this technique really deserves your attention. And because the industry is growing and changing dramatically, the associated costs are dropping.

Another way to use magnets in sleep is to get a high-quality magnetic sleep pad that provides a magnetic field that does not pulse. These generally cost upward of $500 and weigh at least forty pounds. People report sleeping much more soundly and getting deeper and more restorative sleep from well-made magnetic sleep pads, and the effect here is due to a mitochondrial boost, as well.

SLEEPING ON A BED OF NAILS?

A Bulletproof Sleep Induction Mat that stimulates acupressure points is a great example of cutting-edge technology based on an ancient practice. When you're trying to fall asleep, your inner Labrador worries about stuff. You're trying to relax while your fight-or-flight response is wondering, *What should I be worrying about next? Are there any threats nearby, or is it safe to go to sleep?* This sucks up energy and keeps you awake. It also keeps your mitochondria from going into repair mode at night like they are supposed to.

When you lie down on the Sleep Induction Mat, hundreds of small plastic spikes activate your acupressure points. For a short period of time it hurts enough for your body to think, "Oh no, I'm going to die from all of these spikes." It's not really that painful, but it's uncomfortable enough to make your body pay attention.

How does this help you sleep? When you control your instincts and stay on the mat instead of jumping up and running away, your nervous system soon gives up and calms down. This allows relaxation to flood your body and lets you drop into a parasympathetic state. It helps you get into a deep sleep faster than you normally would and increases blood circulation and endorphin release so you can sleep better throughout the night. And your mitochondria are listening—when you are in parasympathetic mode, they go to work on repair. This sleep hack is so powerful that I created a Sleep Induction Mat with longer spikes to work more effectively.

Another way to sleep under pressure is to use air pressure. Some athletes, frequent flyers, and biohackers have started occasionally sleeping in pressurized tubes called hyperbaric oxygen chambers, which increase the air pressure on your body while you breathe pure oxygen. However, even sleeping under extra air pressure without extra oxygen gas still provides a lot more oxygen to your mitochondria, allowing you to get a full night's sleep with amplified mitochondrial function. This is more amplified than the "train high, sleep low" philosophy of athletes who sleep at sea level but work out in low-oxygen, high-altitude environments in an attempt to form more blood cells. Sleeping in a home hyperbaric chamber is like sleeping *below* sea level, where pressure is higher.

This hack probably isn't for you unless you are a high-performance athlete or someone who will spare no expense to give your brain an extra edge. These chambers start at about $5,000, and they are not convenient to use. I have a hyperbaric chamber that I use to increase mitochondrial function, but I don't usually sleep in it. I use it to breathe pure oxygen under pressure for about an hour, which is a great way to recover from a long flight. Even doing that gives me a better night's sleep. If you're going to sleep in one of these, do it with pressurized air with no added oxygen in order to avoid the risk of oxygen toxicity.

Most cities now have facilities you can visit to experience hyperbaric oxygen. It costs somewhere between $50 and $200 a session, and you need twenty sessions minimum to get the full benefits. Again, this might not be necessary for you, but it is absolutely worth a try if you are suffering from substantial mitochondrial dysfunction.

There is one caveat: if you are suffering from an infection from bacteria that thrive on oxygen, you should skip the hyperbaric chamber. Some Lyme disease coinfections prefer lots of oxygen and can get worse with hyperbaric treatments.

YOU ARE WHAT'S AROUND YOU

If you go back one hundred years, there was a huge division in the scientific community at the time. One group of scientists claimed that the body was electric. The other group said that the body was chemical. Long story short, the chemical guys won. That's why we have "big pharma" and so many of our current assumptions about how the body works. But the truth is that our bodies are far more complex than the people on either side of this argument realized.

Yes, the body is chemical. You'll die if you take a chemical called cyanide because it will stop your mitochondria from working. But wait—is that a chemical effect or an electrical effect? Cyanide is a chemical that stops the electrical production of energy in your cells. And that makes it both a chemical and an electrical effect.

As we've come to understand, it's not just one thing that's affecting our bodies—our entire environment affects us. The body responds to chemicals and electricity, sure, but also light, sound, water, and air. And every day we're learning about additional ways that our bodies respond to our environment. In the meantime, we can hack them using the variables that we are already aware of.

STIMULATE YOUR BRAIN

Running a current over your body adds more electrons to your system so that you can make energy more quickly. It sounds crazy, but it's actually one of the simplest ways of hacking the electron transport chain.

There are two options for electrical stimulation. Cranial electrotherapy stimulation (CES) runs an electrical current back and forth across your brain to put your entire brain in the same state while also supercharging the mitochondria. The Russian space program created this technology when faced with the high expense of sending astronauts into space. They decided to cut costs by using fewer astronauts and training them to perform better on less sleep by using CES. Now you, too, can exceed your potential under any circumstances. While I was writing this book, I used the CES form of electrical stimulation on my brain to put myself into a flow state so that I could write better material more quickly.

CES only recently evolved to become a carefully controlled form of brain stimulation called transcranial alternating current stimulation (tACS), where a computer shapes the exact electrical waves that can change your brain activity the most, with more precision than ever before. We use it for brain upgrades at the 40 Years of Zen neurofeedback facility because it can help new neurons grow more quickly and insulate themselves to carry more electrical current. Another less evolved but similar technique is transcranial direct current stimulation (tDCS). You can buy tDCS machines online for $100.

And you don't have to target your brain directly with a current to get the benefits of electrical stimulation. All of the mitochondria in your body talk to each other. That means that if you do electrical stimulation of your muscles, your brain will get some of the benefits. I have used electrical stimulation on my muscles for many years and experienced a profound decrease in muscle pain from previous injuries.

Interestingly, Dr. Terry Wahls, physician and author of *Minding My Mitochondria*, used electrical stimulation along with specific diet recommendations (which are in alignment with Head Strong) to help reverse her mitochondrial dysfunction. And as you read earlier, she was able to stop using her wheelchair as a result, despite having advanced MS. I recommend her book if you're dealing with any chronic neurodegenerative disease.

OZONE THERAPY

Ozone therapy was first developed in Germany in the early 1950s, but it has become popular lately as a radical performance and injury recovery tool—and for good reason. Here's why: Regular oxygen (the kind you breathe) is composed of two oxygen atoms. Ozone, on the other hand, is composed of three oxygen atoms. The addition of this third oxygen atom supercharges oxygen into ozone and gives it powerful healing properties. Ozone regulates the immune system and is used to treat autoimmune diseases as well as cancer, AIDS (acquired immunodeficiency syndrome), and chronic infections. Ozone also stimulates our cells to take up more oxygen. This helps your mitochondria make more energy, and the extra electrons that charge the ozone can enter the Krebs cycle directly, changing the NAD+ to NADH ratio the same way that the strongest mitochondrial supplements like KetoPrime do.

Most important, ozone is a mitochondrial stimulant that protects mitochondria against oxidative stress. A 2015 study found that "[low-level ozone therapy] treatment induces positive and long-lasting cellular responses in cytoskeletal organization and mitochondrial activation."[1]

When I had mitochondrial damage from Lyme disease and toxic mold exposure, I was desperate to get my brain back to full function. I learned how to do ozone therapy from Dr. Timothy Gallagher, DDS, who was a pioneer in the field of ozone dentistry. He was a friend and adviser to the Silicon Valley Health Institute until he passed away in 2014. Because he generously taught me how, I was able to do ozone therapy at home every night for eighteen months. The first night I tried it, I felt my brain turn on after just five minutes, and each night from there it got stronger as my mitochondria recovered. I firmly believe that ozone therapy is one of the main reasons I have better performance today than I did when I was twenty.

There are three types of ozone therapy you can try. The most powerful is called major autohemotherapy (MAH or MAHT), and you do it in a doctor's office. They draw some blood intravenously, mix ozone gas into it, and reintroduce it into your veins. You feel a burst of energy throughout the body as your mitochondria start to heal. The next way of taking ozone is via insufflation, where you take a measured volume of specific-strength ozone gas and put it in your body either vaginally or via the rectum. This is normally done at a doctor's office, but people like me who have dealt with long-term toxin exposure that has harmed their mitochondria can learn to do it at home.

The final type of ozone therapy is very potent for injury recovery. It is called Prolozone, and it can be administered in a doctor's office. With this form of therapy, the physician injects ozone gas and some nutrients into an injury site, such as a knee or the spine. It often takes only a couple of treatments, but it immediately lowers the swelling and turns on healing. I have seen profound effects on knees, elbows, and bulging discs from Prolozone. Anytime you get more oxygen and more mitochondrial function at the same time, good things are going to happen in your body.

Ozone therapy is one of the strongest, most affordable, broadest-spectrum mitochondrial upgrades I've ever seen. If you're dealing with early-onset mitochondrial dysfunction (as 48 percent of people under forty are), or if you're over forty, an occasional ozone treatment has the potential of prolonging your mitochondrial function for years.

CRYOTHERAPY

Are you ready to get cold? You read about all the benefits of cold therapy earlier, and hopefully you've spent at least two weeks either taking cold showers or dunking your face in ice water, or both. But now it's time to turn it up a notch. Find a whole-body cryotherapy center near you, where you can get your cold on in a safe and controlled environment.

With cryotherapy, you stand in super-chilled air that is –270°F for up to three minutes. It's not as unpleasant as it sounds because you are standing in cold air that only chills your outer skin; a cool swimming pool is much more uncomfortable because it chills you to the bone. There is no stronger mitochondrial recharge I can think of that only takes three minutes.

I expect cryotherapy to continue to grow in popularity around the world because it works so well. As you read earlier, a positive side effect of cryotherapy is that it stimulates collagen synthesis in your skin, which means fewer wrinkles and skin that heals better. Since I installed a cryotherapy chamber with liquid nitrogen in my biohacking laboratory, I've received a lot of comments about how my skin looks. The cold helps! And so does eating a lot of the collagen protein I created so that my body has the building blocks to grow healthy skin. Where does the energy to grow collagen come from? You guessed it . . . mitochondria!

LIGHT IT UP

To get more quality light and extreme detoxification benefits, try out a far infrared sauna. It can transform the water in your cells into EZ water so your mitochondria work more efficiently. You can find far infrared saunas at a lot of day spas and yoga studios and gyms, or you can buy one for your own home. The prices on infrared saunas have come way down over the last few years, so buying one for home use is no longer a crazy idea. The amount of time you should spend in an infrared sauna is personalized based on your fitness level and many other factors. Start with twenty to thirty minutes two or three times a week and see how you feel. You might need more or less. I sit in mine for an hour on the highest setting, but that's because I'm used to it! Start with twenty minutes and slowly work your way up.

I also use infrared LEDs and red LEDs on my entire body because they recharge mitochondria more effectively than a sauna, with almost no sweating. The device I use is called the REDcharger and contains more than forty thousand red and infrared LEDs and a very small number of narrow-spectrum blue LEDs at a spectrum documented to reduce wrinkles. I do twenty minutes a day for whole-body mitochondria and skin benefits. You can find a local facility with a REDcharger or get a smaller handheld infrared and red LED device for stimulating collagen and mitochondria. There are even a few red and infrared devices designed to help with thinning hair—mitochondrial activation can save dying hair follicles! The smaller lights require a lot more time for full body effects, so go with more LEDs and more power when you can.

SHAKE, BOUNCE, AND VIBRATE

In addition to your regular Head Strong workout routine, try jumping on a mini trampoline for five to ten minutes every day. This will shake all of the water in your cells, which can increase your levels of EZ water. You'll experience less inflammation, more mental clarity, and a big boost in energy. Plus, it's fun! There are various forms of tai chi and yoga with whole-body shaking exercises that can have a similar effect.

With a mini trampoline, you can get about one bounce per second. I use the Bulletproof Vibe, a mechanically assisted whole-body vibration platform that gives me thirty bounces per second for faster water structuring and mitochondrial activation. I keep one next to my standing desk and often use it for a minute or two between meetings. The difference in whole-body energy is hard to describe, but it is very noticeable. I also make the Bulletproof Vibe available at all of the Bulletproof Coffee shop locations for visitors to experience as we prepare their Bulletproof Coffee. Surprisingly, whole-body vibration isn't new. Nikola Tesla actually pioneered it in his laboratory about a hundred years ago. It's a technology that works well, but most people just don't know about it.

GET GROUNDED

When your mitochondria finish using electrons, the "used" electrons normally attach to oxygen molecules. However, extra electrons can flow to any electrical ground, including the earth itself. That's why you feel more energized when you walk barefoot on grass or on the beach, and it's why electrically earthing yourself can reduce inflammation, improve sleep, and fix jet lag.

To get the benefits, you can walk outside barefoot for a few minutes every day, or you can get a conductive mat for your desk and conductive sheets for your bed. The mat and sheets plug into the little round grounding part of American electrical plugs so electrons can flow from your body out to the electrical system. This isn't as natural as walking barefoot, but it works. A few companies make special grounding shoes, too. At the Bulletproof Coffee shop, we build electrical grounding into the furniture so that people who stop by for coffee can get a grounding mitochondrial upgrade from the furniture.

BEYOND SUPPLEMENTS

The Head Strong recommended supplements can totally change your brain and boost your number of electrons, and they do it with natural substances. There are additional specific mitochondrial stimulants in this section that can improve your performance even more dramatically. Be warned, though, that some of these are seen as rather extreme hacks.

NICOTINE MICRODOSING

People often confuse the effects of caffeine with the effects of coffee, even though caffeine and coffee are different substances. (Coffee has hundreds of chemicals; caffeine is just one of them.) Likewise, nicotine gets the headlines when it comes to tobacco, even though nicotine is not tobacco—it's just one of more than five thousand chemicals in cigarettes. And there's a lot more to nicotine than smoking and addiction. Like caffeine, it's a powerful nootropic "smart drug," and when you get it at low doses in its pure form—without toxins and carcinogens wrapped around it and rolled into a burning cigarette—nicotine can be a formidable (if occasional) biohack with direct effects on your mitochondrial function and a low risk of addiction.

I have never smoked because burning stuff and putting it in your lungs with carbon monoxide is bad for inflammation, bad for mitochondria, and bad for you. But based on experience with pure nicotine, I predict that over the next few years it will become much more popular for performance and cognitive enhancement. After all, about 99 percent of the great works of literature in the last two hundred years ($p > .05$) were written under the influence of caffeine and nicotine, Mother Nature's original smart drugs.

Like caffeine, nicotine is a defense mechanism made by plants to keep from being eaten by animals, bugs, or fungus—in fact, caffeine and nicotine are in the same chemical family. Many plants produce nicotine and store it in their leaves; it's bitter and toxic in large doses, which helps keep hungry animals at bay. Nicotine is most famously found in tobacco, but you'll also find small amounts of it in members of the nightshade family of vegetables—tomatoes, potatoes, and eggplants, for example. There's even a tiny bit of nicotine in cauliflower.

While nicotine is toxic to smaller animals and insects, humans can withstand a good deal of it—and derive benefits from it. When nicotine reaches your brain, it binds to nicotinic receptors (guess where they got their name?), activating pathways that control attention, memory, motor function, and pleasure.

Excess nicotine can be toxic for humans, so go with the lowest dose you can feel, and stick to it. The exact amount that qualifies as "too much" varies widely from person to person based on tolerance. A heavy smoker might be able to tolerate 100 mg/day, but that amount could kill me. That's why I recommend using the lowest possible dose you can feel, starting at 1 mg/day.

When you get the right amount, nicotine does a lot for you. For starters, it gives you faster, more precise motor function. People show more controlled and fluent handwriting after taking nicotine[2] and they're also able to tap their fingers faster on a keyboard without sacrificing accuracy.[3] Nicotine makes you more vigilant, too. Participants who used nicotine patches were able to pay attention to a mentally tiring task longer than controls could.[4] Nicotine gum had the same effect.[5] Nicotine also sharpens your short-term memory: people who took nicotine better recalled a list of words they'd just read and also made fewer mistakes than people given a placebo when repeating a story word for word.[6] Again, the boost in memory came from both patches and gum. And it has been shown that nicotine can even increase synaptic plasticity.[7]

Of course, there are some real downsides to nicotine, the most infamous of which is that it's addictive. Nicotine activates your mesolimbic dopamine system, which scientists have aptly nicknamed the brain's "pleasure pathway." The pleasure pathway is a double-edged sword. Food, sex, love, and certain drugs all cause this part of your brain to light up, sending a euphoric rush of dopamine through your system and leaving you in bliss. If you indulge this system on a regular basis, though, the constant stimulation dulls the pathway. Your receptors start to pull back into your neurons, where they are very hard to activate, and you start to feel physically ill unless you get more of whatever you were enjoying (or something else equally stimulating). This is how dependence starts.

In 2007, a hallmark addiction study ranked twenty common recreational drugs on a scale of 0 to 3, with higher scores indicating a greater risk of dependence. Tobacco clocked in as the third most addictive drug overall. It had a score of 2.21, beaten only by cocaine (2.39) and heroin (3.00).[8]

However, it's important to note that the people in the study were smoking cigarettes, which deliver a substantial 8–20 mg dose of nicotine. The large dose of nicotine lights up your pleasure pathway like a Christmas tree, and it's strongly associated with the sucking behavior that accompanies smoking a cigarette, which creates an addictive association. Other forms of nicotine are different. Nicotine gum, for example, releases only 2–4 mg over the course of twenty to thirty minutes, so you don't get an addictive, euphoric rush from it, but you still get nicotine's benefits. The nicotine spray I prefer delivers a 1 mg dose—about 5 percent of the amount in a cigarette.

Nicotine has a few other pitfalls. Nicotine by itself (separate from tobacco) was associated with cancer in rats[9] and mice.[10] However, the cancer link has never shown up in human studies, and a recent literature review found that there was no evidence to show that it caused cancer in humans. We do know, though, that nicotine is poisonous at high doses. You can get really sick if you overuse it. Nicotine gum, lozenges, or leftover patches could hurt or even kill a pet or a child. Store and treat all forms of nicotine with care.

These downsides are the reason I recommend that, if you decide to use nicotine for your mitochondrial function, you do it at a very low dose (1–2 mg) and only occasionally, as needed. I've tried out many different forms of nicotine, but it's certainly not something I use regularly. I do find, though, that it's useful for an occasional boost. It's amazing before I go onstage for a big keynote speech or before I get interviewed on TV. (Yes, I stack my mitochondrial enhancers before big events!)

You have eight options if you want to occasionally use nicotine as a nootropic, but I would flat-out reject four of those options (this is not an endorsement to start smoking or get addicted—please be careful!). You can smoke, chew, or snort tobacco, use nicotine gum, spray, patches, or lozenges, or smoke the new electronic cigarettes (e-cigs). Smoking, chewing, and snorting tobacco cause cancer, so avoid those. E-cigs (and vaping) are controversial. Some people say they're safe, but I have real concerns about the nanoparticles of heavy metals from the e-cig combustion chambers. You don't want to breathe that stuff in! I tried a high-end e-cig and it caused throat irritation and made me cough even after attempting to get used to it. I don't use or recommend them, especially because they have an oral sensation like smoking that makes them more addictive. (They're still far better than smoking or chewing tobacco, however.)

As I mentioned, nicotine gum releases only 2–4 mg of nicotine over the course of twenty to thirty minutes. You don't get a euphoric rush from it, but you still receive nicotine's energy benefits. Addiction to nicotine gum is possible but not commonly reported. The problem with nicotine gum is that chewing gum causes your trigeminal nerve (which is associated with chewing) to fire more than it should. Save your chewing for eating, and your jaw (and nervous system) will be healthier. Also, every brand of gum I've found has aspartame in it, often along with other questionable artificial sweeteners. Aspartame is an excitatory neurotoxin—avoid it!

Nicotine patches are somewhere in between gum and cigarettes. They contain more nicotine than the gum, but since you absorb it slowly through your skin throughout the day, you get sustained focus and energy. When I tried nicotine patches, I'd take the smallest-dose patch I could find and cut it in half (even though it says not to on the label). I'd leave it on for one to two hours, so I would get 1–4 mg of nicotine during that time.

Nicotine inhalers are relatively hard to find, but Nicorette makes them, and they have no chemicals at all. It's just a sponge with nicotine and a little plastic straw that you suck through to get nicotine-scented air. I like these because they're free of nasty chemicals, but the downside is that the act of sucking on something appears to be addictive. I found myself wanting to take a puff from one when I was sitting at my desk, even when I didn't need or want the energy from it—so I quit!

Nicotine lozenges, like nicotine gum, are full of crappy chemicals and sweeteners such as aspartame, acesulfame potassium (ace-K), and sucralose. The safest one I've found is the Nicorette mini lozenge, which is very small and contains no aspartame. You do get a small dose of unsafe sweetener, but it's so tiny that it's unlikely to matter much. When I take half of the smallest, 2 mg lozenge, I feel a cognitive shift in about fifteen minutes. These lozenges are easy to find in the United States. And make sure to get the mini lozenges, as the large Nicorette lozenges are full of chemicals you don't want to put in your body.

Nicotine spray is a more recent invention. Each spray of 1 mg of nicotine contains vanishingly small amounts of sucralose. You spray it under your tongue and feel it quickly, making it an excellent option when you want a burst of sustained energy. I've done more than one interview while on this, and I find it's great for jet lag or when you have a heavy day ahead of you and want to maintain focus.

If you do decide to try nicotine, treat it carefully. A safe bet would be to take it on an ad hoc basis. Use it if you want to be extra-sharp for a big presentation or a three-hour meeting, but don't use it daily.

METHYLENE BLUE

Before its use as a smart drug in antiaging circles, the chemical methylene blue was used as a dye during diagnostic tests. Scientists found that the blue dye increased oxygen flow to different parts of the body, particularly the brain. It can cross the blood-brain barrier and acts as an antioxidant in the brain. It also improves the efficiency of your mitochondria by carrying more electrons into the electron transport chain and increasing your mitochondria's oxygen consumption.

Initial studies on methylene blue as a smart drug are promising, and it is now available in tablet form or via IV. A 2007 study showed that it can increase a cell's life span.[11] In 2011, another study showed that it can delay the effects of dementia after it has been diagnosed.[12] This is significant because most existing Alzheimer's treatments can only prevent the disease before it has been diagnosed. Animal studies also show that methylene blue is a powerful nootropic. Rats that were treated with methylene blue showed improvements in cognition and memory retention.[13] So did humans in another study, where methylene blue was shown to help short-term memory.[14]

Methylene blue definitely helps mitochondrial respiration, and you can feel the difference if you try it.[15] And if something is inhibiting your mitochondria, methylene blue can trap leaking electrons and keep your metabolism going.[16] The problem is that as doses get higher, methylene blue becomes a *pro-oxidant* and can do the opposite of what you'd expect—cause oxidative stress.[17] Larger doses may also harm your gut bacteria, and if you have high blood pressure, it's not a good idea to try this one. It's also really harmful for babies.

After the first study came out in 2007, I started experimenting with methylene blue, with mixed results. The biggest concern is getting good-quality methylene blue, given that most of what's on the market is chemical grade or used to keep fish tanks clean. I found pharmaceutical-grade methylene blue and kept the dose low. The safe range is 1–4 mg per kg of body weight.[18]

At this point, if you're a super athlete and methylene blue is allowed in your sport, it's worth experimenting with. If you have chronic fatigue syndrome or another mitochondrial disorder that is hurting your quality of life, it's worth experimenting with. If you're anyone else, I think it is safer to use some of the other supplements in Head Strong before trying this one . . . but I do advocate using it if it will help you. It helped me, just not as much as I'd hoped.

SMART DRUGS

I've been on national TV several times to talk about using smart drugs to get my MBA while working full-time. Each time the reporters focused on a specific pharmaceutical I used called modafinil, but they did not cover my favorite class of smart drugs, the racetams. So I wanted to be sure to highlight them for you here.

Racetams are some of the best studied and the oldest cognitive-enhancing pharmaceuticals on the planet, and they have very few side effects. The first of this class of very safe drugs is called piracetam, and I took it for a few years before switching to more recently evolved versions. Four different well-constructed studies show that piracetam improves mitochondrial function.[19] In fact, I believe this may be one of its important mechanisms of action, one that is overlooked in most of its descriptions. This is one of the safest pharmaceuticals on the market, and it makes your brain and your mitochondria work better.

I recommend two forms of this drug that are better absorbed. My favorite is called aniracetam, because it is fat soluble and documented to improve your memory. My second favorite is called phenylpiracetam, which gives you quite a lot of energy. Both of these drugs are available online and are virtually ignored in Western medicine because they're off patent. I take 800 mg of aniracetam and 100 mg of phenylpiracetam most mornings, and I can feel the difference in my brain as a result. This entire book was written on higher doses of those two substances (as well as every other supplement in the book!).

TRAIN YOUR EYES AND EARS

Like the information that enters your brain through your eyes, information that enters your brain through your ears also has an impact on energy in your brain. Hearing is not quite as draining as visual processing because it requires less energy. That's why your ears don't have nearly as many mitochondria as your eyes. But processing a great deal of auditory information does still require some of your brain's energy. So once your mitochondria are working like rock stars, auditory processing is a great area to focus on for an extra energy boost.

When I was in my thirties I experienced a lot of brain drain from auditory stress. Whenever I was in a noisy environment, I got really tired and was less able to focus. I went through a special ear training called auditory integration training (AIT) to fix that. This type of training is not well known, but it is very effective for improving your hearing and your brain energy if sound processing makes you tired.

For this training, I went to my audiologist's office, where she tested each of my ears to see how good they were at hearing all frequencies. It turned out that I was able to hear both very high and very low frequencies, but I had some hearing gaps in the middle. This was something I was unaware of. The gaps weren't big enough to prevent me from passing any standard hearing test with flying colors, but they left me tired and foggy when my brain struggled to process gaps in information that I simply couldn't hear.

You can think of this as a road with potholes. Is a road with a few potholes still paved? Yes. Can you drive over it without much of a problem? Sure. But it's still stressful to drive over those potholes day after day. And the AIT found the potholes in my hearing. First, I had to listen to music that consisted of only potholes. I was listening to all of the things my brain couldn't process. And initially it sounded awful because my brain was working very hard to try to fill in those potholes.

Thanks to the brain's ability to grow and change, little by little my brain filled in those gaps and the potholes were smoothed out. Now I don't get abnormally tired in noisy environments because my brain isn't struggling to process certain frequencies and sucking up energy that I'd rather use to do something meaningful. This is not really a mitochondrial hack—it's just a way of using less energy than you normally would so you can use that extra energy to kick more ass. But it deserves your attention if you care about your brain function as much as I do, and you'll grow neurons faster during the training if you use the other techniques in this book.

At your audiologist's office, AIT will cost you about $500, but there are cheaper versions that you can try at home. The Tomatis Method uses specifically designed sound tracks that are like weight lifting for your ears—and your brain. They go very quickly from high to low frequencies, forcing your ears to listen and your brain to work hard to process a great deal of information efficiently. There are also apps you can download that use a similar method of training.

Fair warning: this training is not fun or relaxing. It is training in the true sense, meaning it's hard work, but it does make you stronger. I highly recommend using one of these training programs after completing the Head Strong program. First, it's important to use the two-week program to turn up your energy production and get rid of the toxins that are sucking your brain energy. Once you complete the two-week program and get a sense of how capable you are as a human being when your energy systems are running efficiently, it's a good time to use some of that extra energy to get even more Head Strong with auditory training.

Likewise, if you want to save brain energy by improving your visual processing system, visit a local Irlen practitioner and get fitted for a custom pair of Irlen lenses that will filter out the frequencies of light that your brain doesn't like. This will save you a tremendous amount of brain energy and drastically improve your performance. For a less expensive option, buy a pair of TrueDark glasses with performance lenses that filter out the most common irritating frequencies.

STEM CELLS—THE FUTURE OF BIOHACKING?

Stem cells are special cells that have the potential to turn into many other types of cells. (Modern treatments with stem cells have nothing to do with fetal tissue!) While I was writing this book, I went to the U.S. Stem Cell Clinic in Sunrise, Florida, where they took my own stem cells from my fat (specifically from my butt, a rich source of stem cells), put them in a centrifuge, and then injected them into my cerebral spinal fluid, where they eventually traveled to my brain. Yes, that officially makes me a butt head.

This is a cutting-edge treatment for multiple sclerosis and other mitochondrial-based degenerative diseases. I don't have MS, but I did experience a concussion about three months before this treatment, when I got such a bad case of food poisoning that I passed out and hit my head on the floor. I know, very elegant. As you read earlier, I also had damaged mitochondria from years of toxic mold exposure. With this treatment, I knew that when the stem cells reached my brain, they would go to the areas where they were most needed and help me grow new brain cells.

This entire process of growing new brain cells takes about six months, so during that time I followed the Head Strong program while doing extra neurofeedback and other hacks to grow the strongest, most efficient new brain cells possible. Unfortunately for me (but fortunately for you), I stayed up late many nights writing this book when I should have been getting extra sleep to help the stem cells—but even then I saw results.

This was not the first time I was injected with stem cells. A year ago, I did a round of stem cell injections to restore old injuries and avoid aging. That time, they took stem cells from my bone marrow and injected them into a shoulder injury, knee injury, spinal soreness, back, face, hair, and . . . the male organ. Without going into too much detail, I'll just say that the effects of stem cell treatment are nothing short of life changing. Stem cells reduce inflammation and turn on healing properties that use mitochondrial energy.

We're still learning more about how we can use stem cells to affect our mitochondria. We recently learned that the type of mitochondrial DNA (mtDNA) that you inherited from your mother, grandmother, and all of the women before them—and the mutations it has gone through since then—profoundly influences how your brain works, as well as all sorts of other parameters of aging. In other words, if you're suffering from a degenerative disease, it may not be your genome for your DNA that is the problem—it may be the mtDNA, the instructions for making the power plants in your cells.[20]

If you get stem cells from another person with different mitochondrial DNA, new cells grown from those stem cells will have different mitochondrial DNA and different performance characteristics. That is amazing. Remember, this is unrelated to the old technology of fetal stem cells; we can now get stem cells from adults and dedifferentiate them.

There are still risks today of receiving stem cells from someone else, because they might turn into the wrong type of tissue. No one wants a toenail growing out of her forehead! But I have no doubt that we will learn to mitigate these risks and use this technology successfully in the near future.

It is my plan to be one of the first humans to receive a mitochondrial transplant, to be able to literally gift my body with a variety of mitochondrial capabilities that I don't have today. That would create profound resilience . . . if it works. And it may be a long time. Hopefully, thanks to the hacks in this book, you and I will still be alive when that day comes, and we'll still be Head Strong.

AFTERWORD

I'll be honest with you. After all of these years as a professional biohacker—spending hundreds of thousands of dollars hacking my biology and traveling to the ends of the world to seek out the most cutting-edge, extreme, and ancient hacks for improving human performance, I was surprised to learn that it really all comes down to our mitochondria. I never would have imagined the extent to which these billions of tiny bacteria that live in all of us are calling the shots, controlling our energy, brains, and performance and basically determining who we are.

It's shocking, but it's also exciting—because while our mitochondria are controlling us, it's entirely possible to take control over them by changing the environment around us. Using the relatively simple techniques that you now know, you can start calling the shots. You can decide how much energy you want to have, how you want to feel, and how you want to treat the people around you.

Since I learned to hack my mitochondria, I've become more successful, sure, but I've also become a better person. I'm more patient. I'm kinder. I have more empathy. And I'm more present when I'm with the people I love—all because I've taken control of my mitochondria, and that has made it easier for me to focus on self-improvement.

You now understand your body on a deeper level than ever before. You know what you can do to help your mitochondria work more efficiently and what you can do to slow them down. Every time you turn on a light, take a bite of food, or move your body, you are deciding how to treat your mitochondria and how much power you want to have.

Once you experience what it's like to have control over your mitochondria instead of vice versa, it's hard to ever go back to the way things used to be. I hope you never do. Instead, I hope you enjoy owning this knowledge and the power it gives you over your own destiny. And I hope you decide to do great things with it.

WHAT'S NEXT?

Go deeper with free online *Head Strong* videos and additional resources. Visit www.bulletproof.com/headstrong for exclusive content, an invitation to join the *Head Strong* reader community, and quick-start study guides to help you wrap your head around everything in the book—whether you learn best by reading, watching, or listening.

Up to the time of printing, it has been hard to get reliable information about how your mitochondria are doing, especially over time. After researching every digital health company out there, I've joined the scientific advisory board of Viome, a truly disruptive wellness company that makes personalized recommendations based on regular comprehensive analyses of your gut microbial reactome, mitochondrial function, and their interactions. Head on over to Viome.com/headstrong to learn more and sign up to get your own data.

Mitochondrial function testing is one of the most important things you can do, and new information is coming on the market all the time. You can access the frequently updated list of mitochondrial tests I recommend at bulletproof.com/headstrong.

ACKNOWLEDGMENTS

The motivation to write *Head Strong* came directly from the millions of Bulletproof followers, readers, and listeners who inspire me daily with their questions and comments on Facebook, the Bulletproof website, and on *Bulletproof Radio*. I'm grateful for your time and attention, and promise to fill every minute of it with the very best information that will help you the most. Knowing this work is making such a huge difference inspires me and makes it fun.

In the middle of writing this book, I suffered a traumatic brain injury that really took me out of my game. I used Head Strong techniques to minimize damage and protect my mitochondria, but they alone weren't enough. Fortunately, I was opening the world's highest-end brain-training center. Special thanks to Dr. Drew Pierson, Jenna and Chris Keane, and the entire team at the 40 Years of Zen neurofeedback brain-training institute for putting my brain all the way back together, and for continuing to make it better every time I visit. Without your hard work, I wouldn't have been able to finish the book.

When you get your brain working all the way—and then some—it's still hard to come up with a complete list of people who have helped you. Should I list all 350 guests who shared their knowledge and experience on *Bulletproof Radio*? The 200 lecturers and presenters who graced the audience at the nonprofit Silicon Valley Health Institute (SVHI) with their knowledge? The entire American Academy of Anti-Aging Medicine because the physicians there have spent twenty years pushing mitochondrial boundaries?

The short answer is yes, and then some. So thank you.

Dr. Barry Morguelon, thanks for writing the mitochondrial meditation for this book, and for the amazing energy your program and treatments have produced for me. Vishen Lakhiani, founder of Mindvalley, has been really helpful with book launch plans. And thanks to Jenna and Chris Keane for running the quarterly Biohacked curated box, which has so many biohacking toys at a discount for true fans.

Special thanks to Barry Sears, author of the famous *Zone Diet,* who really helped me double down on polyphenols in his interview on my show. Dr. David Perlmutter's many works on the brain have helped countless people think better over the years, and Dr. Mark Hyman of the Cleveland Clinic and Dr. David Ludwig of Harvard and Boston Children's have helped shift the national conversation about fat. Dr. Daniel Amen's pioneering work showed me that brain dysfunction isn't a moral failing; it's a hardware failure. Dominic D'Agostino and Dr. Richard Veech are two leading ketone researchers who came on *Bulletproof Radio* to share their knowledge. And eternal thanks to Steve Fowkes, author of one of the first books about smart drugs, a book that saved my cognitive function and career years ago, when my mitochondria were not working. Dr. Frank Shallenberger was the first person to publicly sound the alarm bell about early-onset mitochondrial dysfunction (and show how ozone fixes it)—and for both of these I'm grateful.

Then there are so many people who have helped me learn to effectively share important work using business. Jay Abraham, famous marketing guru, has spent countless hours supporting and mentoring me. Thanks, Jay! Likewise, Joe Polish and his Genius Network have opened doors I'd never imagined were there. Dan Sullivan of Strategic Coach has helped me learn to focus on what matters most, and Cameron Herald has helped me learn how to operate it. My dear friends JJ Virgin and Mike Koenigs have helped me learn how to be a successful author and marketer. I'm inspired every time I get to spend time with the amazing Brendon Burchard. Peter Diamandis and Rick Rubin have each in his own way inspired me to think differently and more deeply. Thanks, guys.

Celeste Fine, my literary agent, must have the most incredible mitochondria in her brain—she's amazing. Jodi Lipper, your lightning-fast editing and writing skills were so helpful. Special thanks to the team at Harper Wave, including Sarah Murphy and Julie Will, editors extraordinaire, and Brian Perrin and Victoria Comella, marketing and publicity rock stars.

Team Bulletproof works tirelessly to help literally millions of people, and I'm grateful beyond words that they have joined in the mission. Thanks for your extra help, Zak Garcia, Susan Lyon, Karen Huh, Amy Herrera, Genevieve Gunderson, Nikki Hoyrup, and Mary Polzella for helping get *Head Strong* written. Steven and Kathleen Crandell, thanks for double-checking and testing all the recipes! Kailey and Gedaly, your work helping to get this book in as many hands as possible is so important. Thank you. Kathleen Raferty and Arthur Page, thanks for the illustrations!

Finally, to my wife, Dr. Lana, and kids, Anna and Alan, thank you for all your patience and support while I stayed up late writing *Head Strong* while still working my day job as CEO of Bulletproof. Dear readers, should you run into my family in person, please thank them for helping *Head Strong* happen! And thanks, Mom, for your mitochondria . . . and Dad, thank you too, but not for your mitochondria! Those were all Mom's. ;)

NOTES

CHAPTER 1: HEAD START

1. Fei Du et al., "Tightly Coupled Brain Activity and Cerebral ATP Metabolic Rate," *Proceedings of the National Academy of Sciences* 105, no. 17 (April 29, 2008): 6409–14, doi:10.1073/pnas.0710766105.
2. Kathleen D. Vohs et al., "Running Head: Self-Regulation and Choice" (Unpublished conference paper, Chicago Booth Marketing Workshop, Chicago, Illinois, 2005), https://www.chicagobooth.edu/research/workshops/marketing/archive/WorkshopPapers/vohs.pdf.

CHAPTER 2: MIGHTY MITOCHONDRIA

1. Carolyn M. Matthews, "Nurturing Your Divine Feminine," *Proceedings (Baylor University Medical Center)* 24, no. 3 (2011): 248.
2. Prakash Seppan et al., "Influence of Testosterone Deprivation on Oxidative Stress Induced Neuronal Damage in Hippocampus of Adult Rats," (Conference poster, 39th American Society of Andrology Annual Meeting, April 6, 2014) *Andrology*, 2 (Suppl. 1) (April 2014): 62, doi:10.1111/j.2047–2927.2014.00221.x.
3. Martyn A. Sharpe, Taylor L. Gist, and David S. Baskin, "Alterations in Sensitivity to Estrogen, Dihydrotestosterone, and Xenogens in B-Lymphocytes from Children with Autism Spectrum Disorder and Their Unaffected Twins/Siblings," *Journal of Toxicology* 2013 (2013).
4. Kathleen A. Mattingly et al., "Estradiol Stimulates Transcription of Nuclear Respiratory Factor-1 and Increases Mitochondrial Biogenesis," *Molecular Endocrinology* 22, no. 3 (March 2008): 609–22, doi:10.1210/me.2007–0029.
5. Yuko Hara et al., "Presynaptic Mitochondrial Morphology in Monkey Prefrontal Cortex Correlates with Working Memory and Is Improved with Estrogen Treatment," *Proceedings of the National Academy of Sciences of the United States of America* 111, no. 1 (January 7, 2014): 486–91, doi:10.1073/pnas.1311310110.
6. Federica Cioffi et al., "Thyroid Hormones and Mitochondria: With a Brief Look at Derivatives and Analogues," *Mitochondrial Endocrinology—*

Mitochondria as Key to Hormones and Metabolism 379, no. 1–2 (October 15, 2013): 51–61, doi:10.1016/j.mce.2013.06.006.

7. Anna Gvozdjáková, *Mitochondrial Medicine: Mitochondrial Metabolism, Diseases, Diagnosis and Therapy* (Springer Science & Business Media, 2008).

CHAPTER 3: BECOME A NEUROMASTER

1. Zu-Hang Sheng, "Mitochondrial Trafficking and Anchoring in Neurons: New Insight and Implications," *Journal of Cell Biology* 204, no. 7 (March 31, 2014): 1087, doi:10.1083/jcb.201312123.
2. Xiao-Hong Zhu et al., "Quantitative Imaging of Energy Expenditure in Human Brain," *Neuroimage* 60, no. 4 (2012): 2107–17.
3. R. Steven Stowers et al., "Axonal Transport of Mitochondria to Synapses Depends on Milton, a Novel *Drosophila* Protein," *Neuron* 36, no. 6 (2002): 1063–77, doi:10.1016/S0896–6273(02)01094–2.; Xiufang Guo et al., "The GTPase dMiro Is Required for Axonal Transport of Mitochondria to *Drosophila* Synapses," *Neuron* 47, no. 3 (2005): 379–93; Huan Ma et al., "KIF5B Motor Adaptor Syntabulin Maintains Synaptic Transmission in Sympathetic Neurons," *Journal of Neuroscience* 29, no. 41 (2009): 13019–29.
4. David G. Nicholls and Samantha L. Budd, "Mitochondria and Neuronal Survival," *Physiological Reviews* 80, no. 1 (2000): 315–60.
5. Zu-Hang Sheng, "Mitochondrial Trafficking and Anchoring in Neurons: New Insight and Implications," *Journal of Cell Biology* 204, no. 7 (March 31, 2014): 1087, doi:10.1083/jcb.201312123.; Robert L. Morris and Peter J. Hollenbeck, "The Regulation of Bidirectional Mitochondrial Transport Is Coordinated with Axonal Outgrowth," *Journal of Cell Science* 104, no. 3 (1993): 917–27; Gordon Ruthel and Peter J. Hollenbeck, "Response of Mitochondrial Traffic to Axon Determination and Differential Branch Growth," *Journal of Neuroscience* 23, no. 24 (2003): 8618–24.
6. Jian-Sheng Kang et al., "Docking of Axonal Mitochondria by Syntaphilin Controls Their Mobility and Affects Short-Term Facilitation," *Cell* 132, no. 1 (2008): 137–48.
7. Zu-Hang Sheng and Qian Cai, "Mitochondrial Transport in Neurons: Impact on Synaptic Homeostasis and Neurodegeneration," *Nature Reviews Neuroscience* 13, no. 2 (2012): 77–93.
8. Sébastien Tremblay et al., "Attentional Filtering of Visual Information by Neuronal Ensembles in the Primate Lateral Prefrontal Cortex," *Neuron* 85, no. 1 (2015): 202–15, doi:10.1016/j.neuron.2014.11.021.
9. A. Lajtha et al., "Turnover of Myelin Proteins in Mouse Brain in Vivo," *Biochemical Journal* 164, no. 2 (May 15, 1977): 323–29.
10. Sidney A. Jones et al., "Triiodothyronine Is a Survival Factor for Developing Oligodendrocytes," *Molecular and Cellular Endocrinology* 199, no. 1–2 (January 31, 2003): 49–60.

11. L. I. Garay et al., "Progesterone Down-Regulates Spinal Cord Inflammatory Mediators and Increases Myelination in Experimental Autoimmune Encephalomyelitis," *Neuroscience* 226 (December 13, 2012): 40–50, doi:10.1016/j.neuroscience.2012.09.032.

12. J. M. Dietschy and S. D. Turley, "Cholesterol Metabolism in the Brain," *Current Opinion in Lipidology* 12, no. 2 (April 2001): 105–12.

13. Stephanie Seneff, Glyn Wainwright, and Luca Mascitelli, "Nutrition and Alzheimer's Disease: The Detrimental Role of a High Carbohydrate Diet," *European Journal of Internal Medicine* 22, no. 2 (n.d.): 134–40, doi:10.1016/j.ejim.2010.12.017.

14. Amy Paturel, "Good Fats—Boost Brain Power with Good Fats," Cleveland Clinic Wellness, September 8, 2009, http://www.clevelandclinicwellness.com/food/GoodFats/Pages/BoostBrainPowerwithGoodFats.aspx.

15. In Young Choi et al., "A Diet Mimicking Fasting Promotes Regeneration and Reduces Autoimmunity and Multiple Sclerosis Symptoms," *Cell Reports* 15, no. 10 (June 7, 2016): 2136–46, doi:10.1016/j.celrep.2016.05.009.

16. A. E. Hoban et al., "Regulation of Prefrontal Cortex Myelination by the Microbiota," *Translational Psychiatry* 6 (April 5, 2016): e774, doi:10.1038/tp.2016.42.

17. "The Life and Death of a Neuron," National Institute of Neurological Disorders and Stroke, July 1, 2015, http://www.ninds.nih.gov/disorders/brain_basics/ninds_neuron.htm.

18. R. Molteni et al., "A High-Fat, Refined Sugar Diet Reduces Hippocampal Brain-Derived Neurotrophic Factor, Neuronal Plasticity, and Learning," *Neuroscience* 112, no. 4 (2002): 803–14.

19. Barbara S. Beltz et al., "Omega-3 Fatty Acids Upregulate Adult Neurogenesis," *Neuroscience Letters* 415, no. 2 (March 26, 2007): 154–58, doi:10.1016/j.neulet.2007.01.010.

20. Yanyan Wang et al., "Green Tea Epigallocatechin-3-Gallate (EGCG) Promotes Neural Progenitor Cell Proliferation and Sonic Hedgehog Pathway Activation during Adult Hippocampal Neurogenesis," *Molecular Nutrition and Food Research* 56, no. 8 (August 2012): 1292–1303, doi:10.1002/mnfr.201200035.

21. Christian Mirescu and Elizabeth Gould, "Stress and Adult Neurogenesis," *Hippocampus* 16, no. 3 (2006): 233–38, doi:10.1002/hipo.20155.

22. Jennifer L. Warner-Schmidt and Ronald S. Duman, "Hippocampal Neurogenesis: Opposing Effects of Stress and Antidepressant Treatment," *Hippocampus* 16, no. 3 (2006): 239–49, doi:10.1002/hipo.20156.

23. "Neurogenesis in Adult Brain: Association with Stress and Depression," *ScienceDaily*, September 2, 2008, https://www.sciencedaily.com/releases/2008/08/080831114717.htm.

24. Miriam S. Nokia et al., "Physical Exercise Increases Adult Hippocampal

Neurogenesis in Male Rats Provided It Is Aerobic and Sustained," *Journal of Physiology* 594, no. 7 (April 1, 2016): 1855–73, doi:10.1113/JP271552.

25. M. S. Kaplan, "Environment Complexity Stimulates Visual Cortex Neurogenesis: Death of a Dogma and a Research Career," *Trends in Neurosciences* 24, no. 10 (October 2001): 617–20.

26. Benedetta Leuner, Erica R. Glasper, and Elizabeth Gould, "Sexual Experience Promotes Adult Neurogenesis in the Hippocampus Despite an Initial Elevation in Stress Hormones," *PLOS ONE* 5, no. 7 (July 14, 2010): e11597, doi:10.1371/journal.pone.0011597.

CHAPTER 4: INFLAMMATION

1. Bharat B. Aggarwal et al., "Inflammation and Cancer: How Hot Is the Link?," *Biochemical Pharmacology* 72, no. 11 (November 30, 2006): 1605–21, doi:10.1016/j.bcp.2006.06.029.

2. Dario Giugliano, Antonio Ceriello, and Katherine Esposito, "The Effects of Diet on Inflammation: Emphasis on the Metabolic Syndrome," *Journal of the American College of Cardiology* 48, no. 4 (August 15, 2006): 677–85, doi:10.1016/j.jacc.2006.03.052.

3. Pritam Das, "Overview—Alzheimer's Disease and Inflammation Lab: Pritam Das–Mayo Clinic Research," Mayo Clinic, accessed October 20, 2016, http://www.mayo.edu/research/labs/alzheimers-disease-inflammation/overview.

4. Arthur A. Simen et al., "Cognitive Dysfunction with Aging and the Role of Inflammation," *Therapeutic Advances in Chronic Disease* 2, no. 3 (May 2011): 175–95, doi:10.1177/2040622311399145.

5. Robin C. Hilsabeck et al., "Cognitive Efficiency Is Associated with Endogenous Cytokine Levels in Patients with Chronic Hepatitis C," *Journal of Neuroimmunology* 221, no. 1–2 (April 2010): 53–61, doi:10.1016/j.jneuroim.2010.01.017.; Tessa N. van den Kommer et al., "The Role of Lipoproteins and Inflammation in Cognitive Decline: Do They Interact?," *Neurobiology of Aging* 33, no. 1 (January 2012): 196.e1–196.e12, doi:10.1016/j.neurobiolaging.2010.05.024.; Shino Magaki et al., "Increased Production of Inflammatory Cytokines in Mild Cognitive Impairment," *Experimental Gerontology* 42, no. 3 (March 2007): 233–40, doi:10.1016/j.exger.2006.09.015.; M. G. Dik et al., "Serum Inflammatory Proteins and Cognitive Decline in Older Persons," *Neurology* 64, no. 8 (April 26, 2005): 1371–77, doi:10.1212/01.WNL.0000158281.08946.68.

6. J. P. Godbout et al., "Exaggerated Neuroinflammation and Sickness Behavior in Aged Mice Following Activation of the Peripheral Innate Immune System," *FASEB Journal: Official Publication of the Federation of American Societies for Experimental Biology* 19, no. 10 (August 2005): 1329–31, doi:10.1096/fj.05–3776fje.; Tomas A. Prolla, "DNA Microarray Analysis of the Aging Brain," *Chemical Senses* 27, no. 3 (March 2002): 299–306.

7. Ryan N. Dilger and Rodney W. Johnson, "Aging, Microglial Cell Priming,

and the Discordant Central Inflammatory Response to Signals from the Peripheral Immune System," *Journal of Leukocyte Biology* 84, no. 4 (October 2008): 932–39, doi:10.1189/jlb.0208108; H. A. Rosczyk, N. L. Sparkman, and R. W. Johnson, "Neuroinflammation and Cognitive Function in Aged Mice Following Minor Surgery," *Experimental Gerontology* 43, no. 9 (September 2008): 840–46, doi:10.1016/j.exger.2008.06.004; Godbout et al., "Exaggerated Neuroinflammation and Sickness Behavior in Aged Mice Following Activation of the Peripheral Innate Immune System," 1329–31; Aine Kelly et al., "Activation of p38 Plays a Pivotal Role in the Inhibitory Effect of Lipopolysaccharide and Interleukin-1 Beta on Long-Term Potentiation in Rat Dentate Gyrus," *Journal of Biological Chemistry* 278, no. 21 (May 23, 2003): 19453–62, doi:10.1074/jbc.M301938200.

8. Arthur A. Simen et al., "Cognitive Dysfunction with Aging and the Role of Inflammation," *Therapeutic Advances in Chronic Disease* 2, no. 3 (May 2011): 175–95, doi:10.1177/2040622311399145.
9. Ibid.
10. L. Å. Hanson, "Immune Effects of the Normal Gut Flora," *Monatsschrift Kinderheilkunde* 146, no. 1 (n.d.): S2–6, doi:10.1007/PL00014761.
11. Roberto Berni Canani et al., "Potential Beneficial Effects of Butyrate in Intestinal and Extraintestinal Diseases," *World Journal of Gastroenterology* 17, no. 12 (March 28, 2011): 1519–28, doi:10.3748/wjg.v17.i12.1519.
12. Matam Vijay-Kumar et al., "Metabolic Syndrome and Altered Gut Microbiota in Mice Lacking Toll-Like Receptor 5," *Science* 328, no. 5975 (April 9, 2010): 228–31, doi:10.1126/science.1179721.
13. Ruth E. Ley et al., "Microbial Ecology: Human Gut Microbes Associated with Obesity," *Nature* 444, no. 7122 (December 21, 2006): 1022–23, doi:10.1038/4441022a.
14. Jean-Pascal De Bandt, Anne-Judith Waligora-Dupriet, and Marie-José Butel, "Intestinal Microbiota in Inflammation and Insulin Resistance: Relevance to Humans," *Current Opinion in Clinical Nutrition and Metabolic Care* 14, no. 4 (July 2011): 334–40, doi:10.1097/MCO.0b013e328347924a.
15. Sergio Davinelli et al., "Enhancement of Mitochondrial Biogenesis with Polyphenols: Combined Effects of Resveratrol and Equol in Human Endothelial Cells," *Immunity and Ageing* 10 (2013): 28, doi:10.1186/1742–4933–10–28.
16. Cristian Sandoval-Acuña, Jorge Ferreira, and Hernán Speisky, "Polyphenols and Mitochondria: An Update on Their Increasingly Emerging ROS-Scavenging Independent Actions," *Archives of Biochemistry and Biophysics* 559 (October 1, 2014): 75–90, doi:10.1016/j.abb.2014.05.017.
17. Antoine Louveau et al., "Structural and Functional Features of Central Nervous System Lymphatic Vessels," *Nature* 523, no. 7560 (July 16, 2015): 337–41, doi:10.1038/nature14432.
18. Carlo Pergola et al., "Testosterone Suppresses Phospholipase D, Causing Sex Differences in Leukotriene Biosynthesis in Human Monocytes,"

FASEB Journal: Official Publication of the Federation of American Societies for Experimental Biology 25, no. 10 (October 2011): 3377–87, doi:10.1096/fj.11–182758.

19. Rainer H. Straub, "The Complex Role of Estrogens in Inflammation," *Endocrine Reviews* 28, no. 5 (December 1, 2006): 521–74, doi:10.1210/er.2007–0001.
20. Anthony J. Harmar et al., "Pharmacology and Functions of Receptors for Vasoactive Intestinal Peptide and Pituitary Adenylate Cyclase-Activating Polypeptide: IUPHAR Review 1," *British Journal of Pharmacology* 166, no. 1 (May 2012): 4–17, doi:10.1111/j.1476–5381.2012.01871.x.
21. Amali E. Samarasinghe, Scott A. Hoselton, and Jane M. Schuh, "Spatio-Temporal Localization of Vasoactive Intestinal Peptide and Neutral Endopeptidase in Allergic Murine Lungs," *Regulatory Peptides* 164, no. 2–3 (September 24, 2010): 151–57, doi:10.1016/j.regpep.2010.05.017.
22. Bronwen Martin et al., "Vasoactive Intestinal Peptide-Null Mice Demonstrate Enhanced Sweet Taste Preference, Dysglycemia, and Reduced Taste Bud Leptin Receptor Expression," *Diabetes* 59, no. 5 (May 2010): 1143–52, doi:10.2337/db09–0807.
23. Mathieu Laplante and David M. Sabatini, "mTOR Signaling in Growth Control and Disease," *Cell* 149, no. 2 (April 13, 2012): 274–93, doi:10.1016/j.cell.2012.03.017.
24. Jacqueline Blundell, Mehreen Kouser, and Craig M. Powell, "Systemic Inhibition of Mammalian Target of Rapamycin Inhibits Fear Memory Reconsolidation," *Neurobiology of Learning and Memory* 90, no. 1 (July 2008): 28–35, doi:10.1016/j.nlm.2007.12.004.
25. Cinzia Dello Russo et al., "Involvement of mTOR Kinase in Cytokine-Dependent Microglial Activation and Cell Proliferation," *Biochemical Pharmacology* 78, no. 9 (November 1, 2009): 1242–51, doi:10.1016/j.bcp.2009.06.097.
26. United States Department of Agriculture, "Profiling Food Consumption in America," in *Agriculture Fact Book, 2001–2002* (Washington, DC: United States Department of Agriculture, Office of Communications, 2003).
27. Alan R. Gaby, "Adverse Effects of Dietary Fructose," *Alternative Medicine Review: A Journal of Clinical Therapeutic* 10, no. 4 (December 2005): 294–306.
28. Sarah Myhill, Norman E. Booth, and John McLaren-Howard, "Targeting Mitochondrial Dysfunction in the Treatment of Myalgic Encephalomyelitis/Chronic Fatigue Syndrome (ME/CFS)—a Clinical Audit," *International Journal of Clinical and Experimental Medicine* 6, no. 1 (2013): 1–15.
29. Douglas C. Wallace, "A Mitochondrial Bioenergetic Etiology of Disease," *Journal of Clinical Investigation* 123, no. 4 (April 2013): 1405–12, doi:10.1172/JCI61398.

30. G. Chevalier et al., "Earthing: Health Implications of Reconnecting the Human Body to the Earth's Surface Electrons," *Journal of Environmental and Public Health*, 2012 (2012):291541, doi:10.1155/2012/291541.

31. Lilach Gavish et al., "Low-Level Laser Irradiation Stimulates Mitochondrial Membrane Potential and Disperses Subnuclear Promyelocytic Leukemia Protein," *Lasers in Surgery and Medicine* 35, no. 5 (2004): 369–76, doi:10.1002/lsm.20108.

32. Cleber Ferraresi, Michael R. Hamblin, and Nivaldo A. Parizotto, "Low-Level Laser (Light) Therapy (LLLT) on Muscle Tissue: Performance, Fatigue and Repair Benefited by the Power of Light," *Photonics and Lasers in Medicine* 1, no. 4 (November 1, 2012): 267–86, doi:10.1515/plm-2012–0032.

CHAPTER 5: BRAIN FUEL

1. Stefano Vendrame et al., "Six-Week Consumption of a Wild Blueberry Powder Drink Increases Bifidobacteria in the Human Gut," *Journal of Agricultural and Food Chemistry* 59, no. 24 (December 28, 2011): 12815–20, doi:10.1021/jf2028686.

2. R. Puupponen-Pimiä et al., "Berry Phenolics Selectively Inhibit the Growth of Intestinal Pathogens," *Journal of Applied Microbiology* 98, no. 4 (April 1, 2005): 991–1000, doi:10.1111/j.1365–2672.2005.02547.x.

3. Theresa E. Cowan et al., "Chronic Coffee Consumption in the Diet-Induced Obese Rat: Impact on Gut Microbiota and Serum Metabolomics," *Journal of Nutritional Biochemistry* 25, no. 4 (April 2014): 489–95, doi:10.1016/j.jnutbio.2013.12.009.

4. Valentina Carito et al., "Effects of Olive Leaf Polyphenols on Male Mouse Brain NGF, BDNF and Their Receptors TrkA, TrkB and p75," *Natural Product Research* 28, no. 22 (2014): 1970–84, doi:10.1080/14786419.2014.918977.

5. Kiyofumi Yamada and Toshitaka Nabeshima, "Brain-Derived Neurotrophic Factor/TrkB Signaling in Memory Processes," *Journal of Pharmacological Sciences* 91, no. 4 (2003): 267–70, doi:10.1254/jphs.91.267.

6. Jeremy P. E. Spencer, "Interactions of Flavonoids and Their Metabolites with Cell Signaling Cascades," in *Nutrigenomics*, ed. Gerald Rimbach, Jürgen Fuchs, and Lester Packer (CRC Press, 2005), 353–78, http://www.crcnetbase.com/doi/abs/10.1201/9781420028096.ch17.

7. Ibid.

8. Massimo D'Archivio et al., "Bioavailability of the Polyphenols: Status and Controversies," *International Journal of Molecular Sciences* 11, no. 4 (March 31, 2010): 1321–42, doi:10.3390/ijms11041321.

9. Jane V. Higdon and Balz Frei, "Coffee and Health: A Review of Recent Human Research," *Critical Reviews in Food Science and Nutrition* 46, no. 2 (2006): 101–23, doi:10.1080/10408390500400009.

10. Kenneth J. Mukamal et al., "Coffee Consumption and Mortality after Acute

Myocardial Infarction: The Stockholm Heart Epidemiology Program," *American Heart Journal* 157, no. 3 (March 2009): 495–501, doi:10.1016/j.ahj.2008.11.009.

11. Harumi Uto-Kondo et al., "Coffee Consumption Enhances High-Density Lipoprotein-Mediated Cholesterol Efflux in Macrophages," *Circulation Research* 106, no. 4 (March 5, 2010): 779–87, doi:10.1161/CIRCRESAHA.109.206615.
12. Yi-Fang Chu et al., "Roasted Coffees High in Lipophilic Antioxidants and Chlorogenic Acid Lactones Are More Neuroprotective Than Green Coffees," *Journal of Agricultural and Food Chemistry* 57, no. 20 (October 28, 2009): 9801–8, doi:10.1021/jf902095z.
13. Esther Lopez-Garcia et al., "The Relationship of Coffee Consumption with Mortality," *Annals of Internal Medicine* 148, no. 12 (June 17, 2008): 904–14.
14. Esther Lopez-Garcia et al., "Coffee Consumption and Risk of Stroke in Women," *Circulation* 119, no. 8 (March 3, 2009): 1116–23, doi:10.1161/CIRCULATIONAHA.108.826164.
15. W. L. Zhang et al., "Coffee Consumption and Risk of Cardiovascular Events and All-Cause Mortality among Women with Type 2 Diabetes," *Diabetologia* 52, no. 5 (May 2009): 810–17, doi:10.1007/s00125–009–1311–1.
16. D. D. Mellor et al., "High-Cocoa Polyphenol-Rich Chocolate Improves HDL Cholesterol in Type 2 Diabetes Patients," *Diabetic Medicine: A Journal of the British Diabetic Association* 27, no. 11 (November 2010): 1318–21.
17. M. Sánchez-Hervás et al., "Mycobiota and Mycotoxin Producing Fungi from Cocoa Beans," *International Journal of Food Microbiology* 125, no. 3 (July 31, 2008): 336–40, doi:10.1016/j.ijfoodmicro.2008.04.021.
18. Mark A. Wilson et al., "Blueberry Polyphenols Increase Lifespan and Thermotolerance in Caenorhabditis Elegans," *Aging Cell* 5, no. 1 (February 2006): 59–68, doi:10.1111/j.1474–9726.2006.00192.x.
19. Ana Rodriguez-Mateos et al., "Intake and Time Dependence of Blueberry Flavonoid-Induced Improvements in Vascular Function: A Randomized, Controlled, Double-Blind, Crossover Intervention Study with Mechanistic Insights into Biological Activity," *American Journal of Clinical Nutrition* 98, no. 5 (November 2013): 1179–91, doi:10.3945/ajcn.113.066639.
20. Navindra P. Seeram, Rupo Lee, and David Heber, "Bioavailability of Ellagic Acid in Human Plasma After Consumption of Ellagitannins from Pomegranate (*Punica Granatum L.*) Juice," Clinica Chimica Acta; *International Journal of Clinical Chemistry* 348, no. 1–2 (October 2004): 63–68, doi:10.1016/j.cccn.2004.04.029
21. Olga Vitseva et al., "Grape Seed and Skin Extracts Inhibit Platelet Function and Release of Reactive Oxygen Intermediates," *Journal of Cardiovascular Pharmacology* 46, no. 4 (October 2005): 445–51.

22. Debasis Bagchi et al., "Molecular Mechanisms of Cardioprotection by a Novel Grape Seed Proanthocyanidin Extract," *Mutation Research* 523–524 (March 2003): 87–97.
23. David Pajuelo et al., "Chronic Dietary Supplementation of Proanthocyanidins Corrects the Mitochondrial Dysfunction of Brown Adipose Tissue Caused by Diet-Induced Obesity in Wistar Rats," *British Journal of Nutrition* 107, no. 2 (January 2012): 170–78, doi:10.1017/S0007114511002728.
24. Junli Zhen et al., "Effects of Grape Seed Proanthocyanidin Extract on Pentylenetetrazole-Induced Kindling and Associated Cognitive Impairment in Rats," *International Journal of Molecular Medicine* 34, no. 2 (August 2014): 391–98, doi:10.3892/ijmm.2014.1796.
25. Valerie Desquiret-Dumas et al., "Resveratrol Induces a Mitochondrial Complex I Dependent Increase in NADH Oxidation Responsible for Sirtuin Activation in Liver Cells," *Journal of Biological Chemistry* (October 31, 2013), doi:10.1074/jbc.M113.466490.
26. Marie Lagouge et al., "Resveratrol Improves Mitochondrial Function and Protects Against Metabolic Disease by Activating SIRT1 and PGC-1 alpha," *Cell* 127, no. 6 (December 15, 2006): 1109–22, doi:10.1016/j.cell.2006.11.013.
27. Richard D. Semba, Luigi Ferrucci, and Benedetta Bartali, "Resveratrol Levels and All-Cause Mortality in Older Community-Dwelling Adults," *JAMA Internal Medicine* 174, no. 7 (July 1, 2014): 1077–84, doi:10.1001/jamainternmed.2014.1582.
28. Tamara Shiner et al., "Dopamine and Performance in a Reinforcement Learning Task: Evidence from Parkinson's Disease," *Brain: A Journal of Neurology* 135, Pt 6 (June 2012): 1871–83, doi:10.1093/brain/aws083.
29. Paul T. Francis et al., "The Cholinergic Hypothesis of Alzheimer's Disease: A Review of Progress," *Journal of Neurology, Neurosurgery and Psychiatry* 66, no. 2 (February 1, 1999): 137–47, doi:10.1136/jnnp.66.2.137.
30. Richard H. Hall, "Neurotransmitters and Sleep" (Lesson outline, Missouri University of Science and Technology, 1998), http://web.mst.edu/~rhall/neuroscience/03_sleep/sleepneuro.pdf.
31. Cecilia Vitali, Cheryl L. Wellington, and Laura Calabresi, "HDL and Cholesterol Handling in the Brain," *Cardiovascular Research* 103, no. 3 (August 1, 2014): 405–13, doi:10.1093/cvr/cvu148.
32. Meharban Singh, "Essential Fatty Acids, DHA and Human Brain," *Indian Journal of Pediatrics* 72, no. 3 (March 2005): 239–42.
33. M. A. Crawford et al., "Evidence for the Unique Function of Docosahexaenoic Acid during the Evolution of the Modern Hominid Brain," *Lipids* 34, no. 1 (1999): S39–S47, doi:10.1007/BF02562227.
34. Karin Yurko-Mauro et al., "Beneficial Effects of Docosahexaenoic Acid on Cognition in Age-Related Cognitive Decline," *Alzheimer's and Dementia: The Journal of the Alzheimer's Association* 6, no. 6 (November 2010): 456–64, doi:10.1016/j.jalz.2010.01.013.

35. Dany Arsenault et al., "DHA Improves Cognition and Prevents Dysfunction of Entorhinal Cortex Neurons in 3xTg-AD Mice," *PLOS ONE* 6, no. 2 (February 23, 2011): e17397, doi:10.1371/journal.pone.0017397.

36. Eric N. Ponnampalam, Neil J. Mann, and Andrew J. Sinclair, "Effect of Feeding Systems on Omega-3 Fatty Acids, Conjugated Linoleic Acid and Trans Fatty Acids in Australian Beef Cuts: Potential Impact on Human Health," *Asia Pacific Journal of Clinical Nutrition* 15, no. 1 (2006): 21–29.

37. J. M. Leheska et al., "Effects of Conventional and Grass-Feeding Systems on the Nutrient Composition of Beef," *Journal of Animal Science* 86, no. 12 (December 2008): 3575–85, doi:10.2527/jas.2007–0565.

38. Gabriela Segura, "Ketogenic Diet—a Connection between Mitochondria and Diet," *DoctorMyhill*, November 20, 2015, http://www.drmyhill.co.uk/wiki/Ketogenic_diet_-_a_connection_between_mitochondria_and_diet.

39. Anssi H. Manninen, "Metabolic Effects of the Very-Low-Carbohydrate Diets: Misunderstood 'Villains' of Human Metabolism," *Journal of the International Society of Sports Nutrition* 1, no. 2 (December 31, 2004): 7–11, doi:10.1186/1550–2783–1–2–7.

40. R. Pasquali et al., "Effect of Dietary Carbohydrates during Hypocaloric Treatment of Obesity on Peripheral Thyroid Hormone Metabolism," *Journal of Endocrinological Investigation* 5, no. 1 (February 1982): 47–52, doi:10.1007/BF03350482.

41. Vigen K. Babayan, "Medium Chain Length Fatty Acid Esters and Their Medical and Nutritional Applications," *Journal of the American Oil Chemists' Society* 58, no. 1 (n.d.): 49A–51A, doi:10.1007/BF02666072.

42. A. A. Gibson et al., "Do Ketogenic Diets Really Suppress Appetite? A Systematic Review and Meta-Analysis," *Obesity Reviews: An Official Journal of the International Association for the Study of Obesity* 16, no. 1 (January 2015): 64–76, doi:10.1111/obr.12230.

43. Mark P. Mattson, Wenzhen Duan, and Zhihong Guo, "Meal Size and Frequency Affect Neuronal Plasticity and Vulnerability to Disease: Cellular and Molecular Mechanisms," *Journal of Neurochemistry* 84, no. 3 (February 2003): 417–31.

CHAPTER 6: BRAIN-INHIBITING FOODS

1. Elvira Larqué et al., "Dietary Trans Fatty Acids Alter the Compositions of Microsomes and Mitochondria and the Activities of Microsome Delta6-Fatty Acid Desaturase and Glucose-6-Phosphatase in Livers of Pregnant Rats," *Journal of Nutrition* 133, no. 8 (August 2003): 2526–31.

2. Wenfeng Yu et al., "Leaky β-Oxidation of a Trans-Fatty Acid: Incomplete β-Oxidation of Elaidic Acid Is Due to the Accumulation of 5-Trans-Tetradecenoyl-Coa and Its Hydrolysis and Conversion to 5-Transtetradecenoylcarnitine in the Matrix of Rat Mitochondria," *Journal of Biological Chemistry* 279, no. 50 (December 10, 2004): 52160–67, doi:10.1074/jbc.M409640200.

3. Dariush Mozaffarian et al., "Dietary Intake of Trans Fatty Acids and Systemic Inflammation in Women," *American Journal of Clinical Nutrition* 79, no. 4 (April 2004): 606–12.

4. Giselle S. Duarte and Adriana Farah, "Effect of Simultaneous Consumption of Milk and Coffee on Chlorogenic Acids' Bioavailability in Humans," *Journal of Agricultural and Food Chemistry* 59, no. 14 (July 27, 2011): 7925–31, doi:10.1021/jf201906p.

5. Zhanguo Gao et al., "Butyrate Improves Insulin Sensitivity and Increases Energy Expenditure in Mice," *Diabetes* 58, no. 7 (July 2009): 1509–17, doi:10.2337/db08-1637.

6. Alessio Fasano, "Zonulin and Its Regulation of Intestinal Barrier Function: The Biological Door to Inflammation, Autoimmunity, and Cancer," *Physiological Reviews* 91, no. 1 (January 2011): 151–75, doi:10.1152/physrev.00003.2008.

7. C. Sategna-Guidetti et al., "Autoimmune Thyroid Diseases and Coeliac Disease," *European Journal of Gastroenterology and Hepatology* 10, no. 11 (November 1998): 927–31.

8. Karen L. Madsen et al., "FK506 Increases Permeability in Rat Intestine by Inhibiting Mitochondrial Function," *Gastroenterology* 109, no. 1 (July 1, 1995): 107–14, doi:10.1016/0016-5085(95)90274-0.

9. Elizabeth A. Novak and Kevin P. Mollen, "Mitochondrial Dysfunction in Inflammatory Bowel Disease," *Frontiers in Cell and Developmental Biology* 3 (2015): 62, doi:10.3389/fcell.2015.00062.

10. Chayma Bouaziz, Hassen Bacha, and Laboratory of Research on Biologically Compatible Compounds, Faculty of Dentistry, Monastir, Tunisia, "Mitochondrial Dysfunctions in Response to Mycotoxins: An Overview," in *Mitochondria: Structure, Functions and Dysfunctions*, ed. Oliver L. Svensson (NOVA Science Publishers, 2011), 811–28, https://www.novapublishers.com/catalog/product_info.php?products_id=46019.

11. Ibid.

12. I. Studer-Rohr et al., "The Occurrence of Ochratoxin A in Coffee," *Food and Chemical Toxicology: An International Journal Published for the British Industrial Biological Research Association* 33, no. 5 (May 1995): 341–55.

13. Y. H. Wei et al., "Effect of Ochratoxin A on Rat Liver Mitochondrial Respiration and Oxidative Phosphorylation," *Toxicology* 36, no. 2–3 (August 1985): 119–30.

14. Herman Meisner, "Energy-Dependent Uptake of Ochratoxin A by Mitochondria," *Archives of Biochemistry and Biophysics* 173, no. 1 (March 1976): 132–40, doi:10.1016/0003-9861(76)90243-5.

15. Yan-Der Hsuuw, Wen-Hsiung Chan, and Jau-Song Yu, "Ochratoxin A Inhibits Mouse Embryonic Development by Activating a Mitochondrion-Dependent Apoptotic Signaling Pathway," *International Journal of Molecular Sciences* 14, no. 1 (January 7, 2013): 935–53, doi:10.3390/ijms14010935.

16. Joseph H. Brewer et al., "Detection of Mycotoxins in Patients with Chronic Fatigue Syndrome," *Toxins* 5, no. 4 (April 11, 2013): 605–17, doi:10.3390/toxins5040605.

17. BIOMIN Holding GmbH, "Biomin Global Mycotoxin Survey 2015," 2015, https://info.biomin.net/acton/fs/blocks/showLandingPage/a/14109/p/p-004e/t/page/fm/17.

18. Diane Benford et al., "Ochratoxin A," *International Programme on Chemical Safety*, WHO Food Additives, Safety Evaluation of Certain Mycotoxins in Food, 74 (2001): 281–415.

19. H. M. Martins, M. M. Guerra, and F. Bernardo, "A Six-Year Survey (1999–2004) of the Ocurrence of Aflatoxin M1 in Daily Products Produced in Portugal," *Mycotoxin Research* 21, no. 3 (September 2005): 192–95, doi:10.1007/BF02959261.

20. M. L. Martins, H. M. Martins, and A. Gimeno, "Incidence of Microflora and of Ochratoxin A in Green Coffee Beans (*Coffea Arabica*)," *Food Additives and Contaminants* 20, no. 12 (December 2003): 1127–31, doi:10.1080/0265203031000l620405.

21. Ibid.

22. Studer-Rohr et al., "The Occurrence of Ochratoxin A in Coffee."

23. Mariano B. M. Ferraz et al., "Kinetics of Ochratoxin A Destruction During Coffee Roasting," *Food Control* 21, no. 6 (June 2010): 872–77, doi:10.1016/j.foodcont.2009.12.001.

24. Rufino Mateo et al., "An Overview of Ochratoxin A in Beer and Wine," *International Journal of Food Microbiology,* Mycotoxins from the Field to the Table, 119, no. 1–2 (October 20, 2007): 79–83, doi:10.1016/j.ijfoodmicro.2007.07.029.

25. Marina V. Copetti et al., "Co-Occurrence of Ochratoxin A and Aflatoxins in Chocolate Marketed in Brazil," *Food Control* 26, no. 1 (July 2012): 36–41, doi:10.1016/j.foodcont.2011.12.023.

26. Saima Majeed et al., "Aflatoxins and Ochratoxin A Contamination in Rice, Corn and Corn Products from Punjab, Pakistan," *Journal of Cereal Science* 58, no. 3 (November 2013): 446–50, doi:10.1016/j.jcs.2013.09.007.

27. Ana-Marija Domijan and Andrey Y. Abramov, "Fumonisin B1 Inhibits Mitochondrial Respiration and Deregulates Calcium Homeostasis—Implication to Mechanism of Cell Toxicity," *International Journal of Biochemistry and Cell Biology* 43, no. 6 (June 2011): 897–904, doi:10.1016/j.biocel.2011.03.003.

28. Puneet Singh et al., "Prolonged Glutamate Excitotoxicity: Effects on Mitochondrial Antioxidants and Antioxidant Enzymes," *Molecular and Cellular Biochemistry* 243, no. 1–2 (January 2003): 139–45.

29. P. Humphries, E. Pretorius, and H. Naude, "Direct and Indirect Cellular Effects of Aspartame on the Brain," *European Journal of Clinical Nutrition* 62, no. 4 (August 8, 2007): 451–62, doi:10.1038/sj.ejcn.1602866.

30. Tamanna Zerin et al., "Effects of Formaldehyde on Mitochondrial Dysfunction and Apoptosis in SK-N-SH Neuroblastoma Cells," *Cell Biology and Toxicology* 31, no. 6 (December 2015): 261–72, doi:10.1007/s10565–015–9309–6.

31. Feng-Yih Yu et al., "Citrinin Induces Apoptosis in HL-60 Cells via Activation of the Mitochondrial Pathway," *Toxicology Letters* 161, no. 2 (February 20, 2006): 143–51, doi:10.1016/j.toxlet.2005.08.009.

32. N. Hauptmann et al., "The Metabolism of Tyramine by Monoamine Oxidase A/B Causes Oxidative Damage to Mitochondrial DNA," *Archives of Biochemistry and Biophysics* 335, no. 2 (November 15, 1996): 295–304, doi:10.1006/abbi.1996.0510.

33. James Hamblin, "The Toxins That Threaten Our Brains," *The Atlantic*, March 18, 2014, http://www.theatlantic.com/health/archive/2014/03/the-toxins-that-threaten-our-brains/284466/.

34. S. Peckham, D. Lowery, and S. Spencer, "Are Fluoride Levels in Drinking Water Associated with Hypothyroidism Prevalence in England? A Large Observational Study of GP Practice Data and Fluoride Levels in Drinking Water," *Journal of Epidemiology and Community Health* 69, no. 7 (July 2015): 619–24, doi:10.1136/jech-2014–204971.

35. Brenda Goodman, "Pesticide Exposure in Womb Linked to Lower IQ," *WebMD*, April 21, 2011, http://www.webmd.com/baby/news/20110421/pesticide-exposure-in-womb-linked-to-lower-iq.

36. Somayyeh Karami-Mohajeri, and Mohammad Abdollahi, "Mitochondrial Dysfunction and Organophosphorus Compounds," *Toxicology and Applied Pharmacology* 270, no. 1 (July 1, 2013): 39–44, doi:10.1016/j.taap.2013.04.001.

37. Alessia Carocci et al., "Mercury Toxicity and Neurodegenerative Effects," *Reviews of Environmental Contamination and Toxicology* 229 (2014): 1–18, doi:10.1007/978–3–319–03777–6_1.

38. James Hamblin, "The Toxins That Threaten Our Brains," *The Atlantic*, March 18, 2014, http://www.theatlantic.com/health/archive/2014/03/the-toxins-that-threaten-our-brains/284466/.

39. Paul K. Crane et al., "Glucose Levels and Risk of Dementia," *New England Journal of Medicine* 369, no. 6 (August 8, 2013): 540–48, doi:10.1056/NEJMoa1215740.

40. Rahul Agrawal and Fernando Gomez-Pinilla, "'Metabolic Syndrome' in the Brain: Deficiency in Omega-3 Fatty Acid Exacerbates Dysfunctions in Insulin Receptor Signalling and Cognition," *Journal of Physiology* 590, no. 10 (May 15, 2012): 2485–99, doi:10.1113/jphysiol.2012.230078.

41. Alan R. Gaby, "Adverse Effects of Dietary Fructose," *Alternative Medicine Review: A Journal of Clinical Therapeutic* 10, no. 4 (December 2005): 294–306.

42. Natasha Jaiswal et al., "Fructose Induces Mitochondrial Dysfunction

and Triggers Apoptosis in Skeletal Muscle Cells by Provoking Oxidative Stress," *Apoptosis: An International Journal on Programmed Cell Death* 20, no. 7 (July 2015): 930–47, doi:10.1007/s10495–015–1128-y.

43. Jan B. Hoek, Alan Cahill, and John G. Pastorino, "Alcohol and Mitochondria: A Dysfunctional Relationship," *Gastroenterology* 122, no. 7 (June 2002): 2049–63, doi:10.1053/gast.2002.33613.

44. Aiden Haghikia, Stefanie Jörg et al., "Dietary Fatty Acids Directly Impact Central Nervous System Autoimmunity via the Small Intestine," *Immunity* 43, no.4 (October 2015): 817–829.

45. Elan D. Louis et al., "Elevated Blood Harmane (1-Methyl-9h-pyrido[3,4-B] indole) Concentrations in Essential Tremor," *Neurotoxicology* 29, no. 2 (March 2008): 294–300, doi:10.1016/j.neuro.2007.12.001.

46. C. D. Davis et al., "Cardiotoxicity of Heterocyclic Amine Food Mutagens in Cultured Myocytes and in Rats," *Toxicology and Applied Pharmacology* 124, no. 2 (February 1994): 201–11.

47. Satoru Takahashi et al., "Chronic Administration of the Mutagenic Heterocyclic Amine 2-Amino-1-Methyl-6-Phenylimidazo[4,5-B]pyridine Induces Cardiac Damage with Characteristic Mitochondrial Changes in Fischer Rats," *Toxicologic Pathology* 24, no. 3 (May 1, 1996): 273–77.

48. Seema Bansal et al., "Mitochondrial Targeting of Cytochrome P450 (CYP) 1B1 and Its Role in Polycyclic Aromatic Hydrocarbon-Induced Mitochondrial Dysfunction," *Journal of Biological Chemistry* 289, no. 14 (April 4, 2014): 9936–51, doi:10.1074/jbc.M113.525659.

49. Ioana Ferecatu et al., "Polycyclic Aromatic Hydrocarbon Components Contribute to the Mitochondria-Antiapoptotic Effect of Fine Particulate Matter on Human Bronchial Epithelial Cells via the Aryl Hydrocarbon Receptor," *Particle and Fibre Toxicology* 7, no. 1 (2010): 18, doi:10.1186/1743–8977–7–18.

50. G. Bounous and P. Gold, "The Biological Activity of Undenatured Dietary Whey Proteins: Role of Glutathione," *Clinical and Investigative Medicine. Médecine Clinique et Experimentale* 14, no. 4 (August 1991): 296–309.

51. Naila Rabbani and Paul J. Thornalley, "Dicarbonyls Linked to Damage in the Powerhouse: Glycation of Mitochondrial Proteins and Oxidative Stress," *Biochemical Society Transactions* 36, Pt 5 (October 2008): 1045–50, doi:10.1042/BST0361045.

52. Pamela Boon Li Pun and Michael P. Murphy, "Pathological Significance of Mitochondrial Glycation," *International Journal of Cell Biology* 2012 (2012): 13, doi:10.1155/2012/843505.

53. Poonamjot Deol et al., "Soybean Oil Is More Obesogenic and Diabetogenic Than Coconut Oil and Fructose in Mouse: Potential Role for the Liver," *PLOS ONE* 10, no. 7 (July 22, 2015): e0132672, doi:10.1371/journal.pone.0132672.

54. Bin Wu et al., "Dietary Corn Oil Promotes Colon Cancer by Inhibiting

Mitochondria-Dependent Apoptosis in Azoxymethane-Treated Rats," *Experimental Biology and Medicine* 229, no. 10 (November 2004): 1017–25.

CHAPTER 7: AVOID TOXINS AND IMPROVE YOUR BODY'S DETOX SYSTEMS

1. Hossam El-Din and M. Omar, "Mycotoxins-Induced Oxidative Stress and Disease," in *Mycotoxin and Food Safety in Developing Countries*, ed. Hussaini Makun (InTech, 2013), http://www.intechopen.com/books/mycotoxin-and-food-safety-in-developing-countries/mycotoxins-induced-oxidative-stress-and-disease.
2. Peter F. Surai et al., "Mycotoxins and Animal Health: From Oxidative Stress to Gene Expression," *Krmiva* 50, no. 1 (March 10, 2008): 35–43.
3. El-Din and Omar, "Mycotoxins-Induced Oxidative Stress and Disease."
4. Kunio Doi and Koji Uetsuka, "Mechanisms of Mycotoxin-Induced Neurotoxicity Through Oxidative Stress-Associated Pathways," *International Journal of Molecular Sciences* 12, no. 8 (August 15, 2011): 5213–37, doi:10.3390/ijms12085213.
5. Elena A. Belyaeva et al., "Mitochondria as an Important Target in Heavy Metal Toxicity in Rat Hepatoma AS-30D Cells," *Toxicology and Applied Pharmacology* 231, no. 1 (August 15, 2008): 34–42, doi:10.1016/j.taap.2008.03.017.
6. Elena A. Belyaeva et al., "Mitochondrial Electron Transport Chain in Heavy Metal–Induced Neurotoxicity: Effects of Cadmium, Mercury, and Copper," *Scientific World Journal* 2012 (April 24, 2012), doi:10.1100/2012/136063.
7. S. Xu et al., "Cadmium Induced Drp1-Dependent Mitochondrial Fragmentation by Disturbing Calcium Homeostasis in Its Hepatotoxicity," *Cell Death and Disease* 4, no. 3 (March 14, 2013): e540, doi:10.1038/cddis.2013.7.
8. C. B. Devi et al., "Developmental Lead Exposure Alters Mitochondrial Monoamine Oxidase and Synaptosomal Catecholamine Levels in Rat Brain," *International Journal of Developmental Neuroscience: The Official Journal of the International Society for Developmental Neuroscience* 23, no. 4 (June 2005): 375–81, doi:10.1016/j.ijdevneu.2004.11.003.
9. A. M. Watrach, "Degeneration of Mitochondria in Lead Poisoning," *Journal of Ultrastructure Research* 10, no. 3 (April 1, 1964): 177–81, doi:10.1016/S0022–5320(64)80001–0.
10. James Dykens, "Drug-Induced Mitochondrial Dysfunction: An Emerging Model for Idiosyncratic Drug Toxicity" (Presentation, MitoAction teleconference, Online, 2009), http://www.mitoaction.org/files/Dykens%20for%20Mitoaction.pdf.
11. Sameer Kalghatgi et al., "Bactericidal Antibiotics Induce Mitochondrial Dysfunction and Oxidative Damage in Mammalian Cells," *Science Translational Medicine* 5, no. 192 (July 3, 2013): 192ra85, doi:10.1126/scitranslmed.3006055.

12. Xu Wang et al., "Antibiotic Use and Abuse: A Threat to Mitochondria and Chloroplasts with Impact on Research, Health, and Environment," *BioEssays* 37, no. 10 (2015): 1045–53, doi:10.1002/bies.201500071.
13. J. L. Stauber and T. M. Florence, "A Comparative Study of Copper, Lead, Cadmium and Zinc in Human Sweat and Blood," *Science of the Total Environment* 74 (August 1, 1988): 235–47, doi:10.1016/0048–9697(88)90140–4.
14. Stephen J. Genuis et al., "Blood, Urine, and Sweat (BUS) Study: Monitoring and Elimination of Bioaccumulated Toxic Elements," *Archives of Environmental Contamination and Toxicology* 61, no. 2 (August 2011): 344–57, doi:10.1007/s00244–010–9611–5.

CHAPTER 8: YOUR BRAIN ON LIGHT, AIR, AND COLD

1. Damian Moran, Rowan Softley, and Eric J. Warrant, "The Energetic Cost of Vision and the Evolution of Eyeless Mexican Cavefish," *Science Advances* 1, no. 8 (September 11, 2015): e1500363–e1500363, doi:10.1126/sciadv.1500363.
2. Martin Picard, "Mitochondrial Synapses: Intracellular Communication and Signal Integration," *Trends in Neurosciences* 38, no. 8 (August 1, 2015): 468–74, doi:10.1016/j.tins.2015.06.001.
3. Bernard F. Godley et al., "Blue Light Induces Mitochondrial DNA Damage and Free Radical Production in Epithelial Cells," *Journal of Biological Chemistry* 280, no. 22 (June 3, 2005): 21061–66, doi:10.1074/jbc.M502194200.
4. Cora Roehlecke et al., "The Influence of Sublethal Blue Light Exposure on Human RPE Cells," *Molecular Vision* 15 (2009): 1929–38.
5. M. A. Mainster, "Light and Macular Degeneration: A Biophysical and Clinical Perspective," *Eye* 1 (Pt 2) (1987): 304–10, doi:10.1038/eye.1987.49.
6. H. R. Taylor et al., "The Long-Term Effects of Visible Light on the Eye," *Archives of Ophthalmology* 110, no. 1 (January 1992): 99–104.
7. T. H. Margrain et al., "Do Blue Light Filters Confer Protection Against Age-Related Macular Degeneration?," *Progress in Retinal and Eye Research* 23, no. 5 (September 2004): 523–31, doi:10.1016/j.preteyeres.2004.05.001.
8. Ronald Klein et al., "The Epidemiology of Retinal Reticular Drusen," *American Journal of Ophthalmology* 145, no. 2 (February 2008): 317–26, doi:10.1016/j.ajo.2007.09.008.
9. Tim Howard, "Colors: Why Isn't the Sky Blue?," Podcast audio, *Radiolab* (WNYC, May 21, 2012), http://www.radiolab.org/story/211213-sky-isnt-blue/.
10. N. A. Rybnikova, A. Haim, and B. A. Portnov, "Does Artificial Light-at-Night Exposure Contribute to the Worldwide Obesity Pandemic?," *International Journal of Obesity* 40, no. 5 (May 2016): 815–23, doi:10.1038/ijo.2015.255.
11. Rosario Rizzuto, "The Collagen-Mitochondria Connection," *Nature Genetics* 35, no. 4 (December 2003): 300–301, doi:10.1038/ng1203–300.

12. Martin Helan et al., "Hypoxia Enhances BDNF Secretion and Signaling in Pulmonary Artery Endothelial Cells" (Unpublished conference paper, American Society of Anesthesiologists, Anesthesiology Annual Meeting, Washington, DC, October 6, 2012), http://www.asaabstracts.com/strands/asaabstracts/abstract.htm;jsessionid=281DD5C69F19839A5616F972343509DF?year=2012&index=9&absnum=3709.
13. Francesco L. Valentino et al., "Measurements and Trend Analysis of O_2, CO_2 and delta13C of CO_2 from the High Altitude Research Station Jungfraujoch, Switzerland—a Comparison with the Observations from the Remote Site Puy de Dôme, France," *Science of the Total Environment* 391, no. 2–3 (March 1, 2008): 203–10, doi:10.1016/j.scitotenv.2007.10.009; C. Sirignano et al., "Atmospheric Oxygen and Carbon Dioxide Observations from Two European Coastal Stations 2000–2005: Continental Influence, Trend Changes and APO Climatology," *Atmospheric Chemistry and Physics* 10, no. 4 (February 15, 2010): 1599–1615, doi:10.5194/acp-10–1599–2010; Y. Tohjima et al., "Gas-Chromatographic Measurements of the Atmospheric Oxygen/Nitrogen Ratio at Hateruma Island and Cape Ochi-Ishi, Japan," *Geophysical Research Letters* 30, no. 12 (June 2003): 1653, doi:10.1029/2003GL017282.
14. C. A. Ramos, H. T. Wolterbeek, and S. M. Almeida, "Exposure to Indoor Air Pollutants during Physical Activity in Fitness Centers," *Building and Environment* 82 (December 2014): 349–60, doi:10.1016/j.buildenv.2014.08.026.
15. Angel A. Zaninovich et al., "Mitochondrial Respiration in Muscle and Liver from Cold-Acclimated Hypothyroid Rats," *Journal of Applied Physiology* 95, no. 4 (October 1, 2003): 1584–90, doi:10.1152/japplphysiol.00363.2003.
16. Véronique Ouellet et al., "Brown Adipose Tissue Oxidative Metabolism Contributes to Energy Expenditure during Acute Cold Exposure in Humans," *Journal of Clinical Investigation* 122, no. 2 (February 1, 2012): 545–52, doi:10.1172/JCI60433.
17. J. Leppäluoto et al., "Effects of Long-Term Whole-Body Cold Exposures on Plasma Concentrations of ACTH, Beta-Endorphin, Cortisol, Catecholamines and Cytokines in Healthy Females," *Scandinavian Journal of Clinical and Laboratory Investigation* 68, no. 2 (2008): 145–53, doi:10.1080/00365510701516350.
18. Anna Lubkowska, Barbara Dołęgowska, and Zbigniew Szyguła, "Whole-Body Cryostimulation—Potential Beneficial Treatment for Improving Antioxidant Capacity in Healthy Men—Significance of the Number of Sessions," *PLOS ONE* 7, no. 10 (October 15, 2012): e46352, doi:10.1371/journal.pone.0046352.
19. Hans-Rudolf Berthoud and Winfried L. Neuhuber, "Functional and Chemical Anatomy of the Afferent Vagal System," *Autonomic Neuroscience*, Fever: The Role of the Vagus Nerve, 85, no. 1–3 (December 20, 2000): 1–17, doi:10.1016/S1566–0702(00)00215–0.
20. Karen L. Teff, "Visceral Nerves: Vagal and Sympathetic Innervation," *Jour-*

nal of Parenteral and Enteral Nutrition 32, no. 5 (October 2008): 569–71, doi:10.1177/0148607108321705.

CHAPTER 9: SLEEP HARDER, MEDITATE FASTER, EXERCISE LESS

1. Lulu Xie et al., "Sleep Drives Metabolite Clearance from the Adult Brain," *Science* 342, no. 6156 (October 18, 2013): 373–77, doi:10.1126/science.1241224.
2. Antoine Louveau et al., "Structural and Functional Features of Central Nervous System Lymphatic Vessels," *Nature* 523, no. 7560 (July 16, 2015): 337–41, doi:10.1038/nature14432.
3. Cristina Carvalho et al., "Cerebrovascular and Mitochondrial Abnormalities in Alzheimer's Disease: A Brief Overview," *Journal of Neural Transmission* 123, no. 2 (January 2015): 107–11, doi:10.1007/s00702–015–1367–7.
4. Xie et al., "Sleep Drives Metabolite Clearance from the Adult Brain."
5. Vaddanahally T. Maddaiah et al., "Effect of Growth Hormone on Mitochondrial Protein Synthesis," *Journal of Biological Chemistry* 248, no. 12 (June 25, 1973): 4263–68.
6. Guang Yang et al., "Sleep Promotes Branch-Specific Formation of Dendritic Spines after Learning," *Science* 344, no. 6188 (June 6, 2014): 1173–78, doi:10.1126/science.1249098.
7. Kristen L Knutson, "Impact of Sleep and Sleep Loss on Glucose Homeostasis and Appetite Regulation," *Sleep Medicine Clinics* 2, no. 2 (June 2007): 187–97, doi:10.1016/j.jsmc.2007.03.004.
8. Laurent Brondel et al., "Acute Partial Sleep Deprivation Increases Food Intake in Healthy Men," *American Journal of Clinical Nutrition* 91, no. 6 (June 2010): 1550–59, doi:10.3945/ajcn.2009.28523.
9. Ryan J. Ramezani and Peter W. Stacpoole, "Sleep Disorders Associated with Primary Mitochondrial Diseases," *Journal of Clinical Sleep Medicine* 10, no. 11 (November 15, 2014): 1233–39, doi:10.5664/jcsm.4212.
10. Wendy M. Troxel et al., "Sleep Symptoms Predict the Development of the Metabolic Syndrome," *Sleep* 33, no. 12 (December 2010): 1633–40.
11. Eileen Luders et al., "The Unique Brain Anatomy of Meditation Practitioners: Alterations in Cortical Gyrification," *Frontiers in Human Neuroscience* 6 (February 29, 2012): 34, doi:10.3389/fnhum.2012.00034.
12. "Brain Gyrification and Its Significance," *Standford VISTALAB Wiki*, June 8, 2013, http://scarlet.stanford.edu/teach/index.php/Brain_Gyrification_and_its_Significance#Relevance_to_Species_Intelligence.
13. "Meditation: In Depth," *NCCIH*, February 1, 2006, https://nccih.nih.gov/health/meditation/overview.htm.
14. Sara W. Lazar et al., "Meditation Experience Is Associated with Increased Cortical Thickness," *Neuroreport* 16, no. 17 (November 28, 2005): 1893–97.
15. Brigid Schulte, "Harvard Neuroscientist: Meditation Not Only Reduces

Stress, Here's How It Changes Your Brain," *Washington Post*, May 26, 2015, https://www.washingtonpost.com/news/inspired-life/wp/2015/05/26/harvard-neuroscientist-meditation-not-only-reduces-stress-it-literally-changes-your-brain/.

16. Huiyun Liang and Walter F. Ward, "PGC-1alpha: A Key Regulator of Energy Metabolism," *Advances in Physiology Education* 30, no. 4 (December 2006): 145–51, doi:10.1152/advan.00052.2006.

17. Martin J. Gibala et al., "Brief Intense Interval Exercise Activates AMPK and p38 MAPK Signaling and Increases the Expression of PGC-1alpha in Human Skeletal Muscle," *Journal of Applied Physiology* 106, no. 3 (March 2009): 929–34, doi:10.1152/japplphysiol.90880.2008.

18. John J. Ratey and Eric Hagerman, *Spark: The Revolutionary New Science of Exercise and the Brain* (Boston: Little, Brown, 2008), http://www.goodreads.com/work/best_book/376155-spark-the-revolutionary-new-science-of-exercise-and-the-brain.

19. Mark P. Mattson, Stuart Maudsley, and Bronwen Martin, "BDNF and 5-HT: A Dynamic Duo in Age-Related Neuronal Plasticity and Neurodegenerative Disorders," *Trends in Neurosciences* 27, no. 10 (October 2004): 589–94, doi:10.1016/j.tins.2004.08.001.

20. Christiane D. Wrann et al., "Exercise Induces Hippocampal BDNF through a PGC-1α/FNDC5 Pathway," *Cell Metabolism* 18, no. 5 (November 5, 2013): 649–59, doi:10.1016/j.cmet.2013.09.008.

21. Kevin T. Gobeske et al., "BMP Signaling Mediates Effects of Exercise on Hippocampal Neurogenesis and Cognition in Mice," *PLOS ONE* 4, no. 10 (October 20, 2009): e7506, doi:10.1371/journal.pone.0007506.

22. J. Eric Ahlskog, "Does Vigorous Exercise Have a Neuroprotective Effect in Parkinson Disease?," *Neurology* 77, no. 3 (July 19, 2011): 288–94, doi:10.1212/WNL.0b013e318225ab66.

23. Olga Khazan, "For Depression, Prescribing Exercise Before Medication," *The Atlantic*, March 24, 2014, http://www.theatlantic.com/health/archive/2014/03/for-depression-prescribing-exercise-before-medication/284587/.

24. Maggie Morehart, "BDNF Basics: 7 Ways to Train Your Brain," *Breaking Muscle*, accessed October 27, 2016, https://breakingmuscle.com/health-medicine/bdnf-basics-7-ways-to-train-your-brain.

25. Kirk I. Erickson et al., "Exercise Training Increases Size of Hippocampus and Improves Memory," *Proceedings of the National Academy of Sciences* 108, no. 7 (February 15, 2011): 3017–22.

26. Neha Gothe et al., "The Acute Effects of Yoga on Executive Function," *Journal of Physical Activity and Health* 10, no. 4 (May 2013): 488–95.

27. V. R. Hariprasad et al., "Yoga Increases the Volume of the Hippocampus in Elderly Subjects," *Indian Journal of Psychiatry* 55, Suppl. 3 (July 2013): S394–96, doi:10.4103/0019–5545.116309.

28. Pamela Byrne Schiller, *Start Smart!: Building Brain Power in the Early Years* (Beltsville, MD: Gryphon House, 1999).
29. Paul Dennison, *Switching On: The Whole Brain Answer to Dyslexia* (Edu-Kinesthetics, 1981).
30. Per Aagaard et al., "Increased Rate of Force Development and Neural Drive of Human Skeletal Muscle Following Resistance Training," *Journal of Applied Physiology* 93, no. 4 (October 1, 2002): 1318–26, doi:10.1152/japplphysiol.00283.2002.
31. Eino Havas et al., "Lymph Flow Dynamics in Exercising Human Skeletal Muscle as Detected by Scintography," *Journal of Physiology* 504, no. 1 (October 1997): 233–39, doi:10.1111/j.1469–7793.1997.233bf.x.
32. P. J. O'Connor, M. P. Herring, and A. Caravalho, "Mental Health Benefits of Strength Training in Adults," *American Journal of Lifestyle Medicine* 4, no. 5 (September 1, 2010): 377–96, doi:10.1177/1559827610368771.
33. W. Kraemer et al., "Endogenous Anabolic Hormonal and Growth Factor Responses to Heavy Resistance Exercise in Males and Females," *International Journal of Sports Medicine* 12, no. 2 (April 1991): 228–35, doi:10.1055/s-2007–1024673.
34. M. J. Schaaf et al., "Circadian Variation in BDNF mRNA Expression in the Rat Hippocampus," *Molecular Brain Research* 75, no. 2 (February 22, 2000): 342–44.
35. Joshua F. Yarrow et al., "Training Augments Resistance Exercise Induced Elevation of Circulating Brain Derived Neurotrophic Factor (BDNF)," *Neuroscience Letters* 479, no. 2 (July 2010): 161–65, doi:10.1016/j.neulet.2010.05.058.
36. Thomas Seifert et al., "Endurance Training Enhances BDNF Release from the Human Brain," *American Journal of Physiology—Regulatory, Integrative and Comparative Physiology* 298, no. 2 (February 1, 2010): R372–77, doi:10.1152/ajpregu.00525.2009.
37. Roy J. Shephard, "Absolute versus Relative Intensity of Physical Activity in a Dose-Response Context:," *Medicine and Science in Sports and Exercise* 33, Suppl. (June 2001): S400–418, doi:10.1097/00005768–200106001–00008.
38. Hannah Steinberg et al., "Exercise Enhances Creativity Independently of Mood," *British Journal of Sports Medicine* 31, no. 3 (September 1997): 240–45, doi:10.1136/bjsm.31.3.240.
39. Ibid.
40. Stephen H. Boutcher, "High-Intensity Intermittent Exercise and Fat Loss," *Journal of Obesity* 2011 (2011), doi:10.1155/2011/868305.
41. Cinthia Maria Saucedo Marquez et al., "High-Intensity Interval Training Evokes Larger Serum BDNF Levels Compared with Intense Continuous Exercise," *Journal of Applied Physiology* 119, no. 12 (December 15, 2015): 1363–73, doi:10.1152/japplphysiol.00126.2015.

42. William E. Brownell, Feng Qian, and Bahman Anvari, "Cell Membrane Tethers Generate Mechanical Force in Response to Electrical Stimulation," *Biophysical Journal* 99, no. 3 (n.d.): 845–52, doi:10.1016/j.bpj.2010.05.025.

43. Ioana Ferecatu et al., "Polycyclic Aromatic Hydrocarbon Components Contribute to the Mitochondria-Antiapoptotic Effect of Fine Particulate Matter on Human Bronchial Epithelial Cells via the Aryl Hydrocarbon Receptor," *Particle and Fibre Toxicology* 7, no. 1 (2010): 18, doi:10.1186/1743–8977–7–18.

44. Andrei P. Sommer, Mike Kh. Haddad, and Hans-Jörg Fecht, "Light Effect on Water Viscosity: Implication for ATP Biosynthesis," *Scientific Reports* 5 (July 8, 2015): 12029, doi:10.1038/srep12029.

45. Arturo Solis Herrera, "Einstein Cosmological Constant, the Cell, and the Intrinsic Property of Melanin to Split and Re-Form the Water Molecule," *MOJ Cell Science and Report* 1, no. 2 (August 27, 2014), doi:10.15406/mojcsr.2014.01.00011.

46. Ana S. P. Moreira et al., "Coffee Melanoidins: Structures, Mechanisms of Formation and Potential Health Impacts," *Food and Function* 3, no. 9 (September 2012): 903–15, doi:10.1039/c2fo30048f.

CHAPTER 11: HEAD STRONG LIFESTYLE

1. Joshua J. Gooley et al., "Exposure to Room Light Before Bedtime Suppresses Melatonin Onset and Shortens Melatonin Duration in Humans," *Journal of Clinical Endocrinology and Metabolism* 96, no. 3 (March 2011): E463–72, doi:10.1210/jc.2010–2098.

2. Joshua J. Gooley et al., "Spectral Responses of the Human Circadian System Depend on the Irradiance and Duration of Exposure to Light," *Science Translational Medicine* 2, no. 31 (May 12, 2010): 31ra33–31ra33, doi:10.1126/scitranslmed.3000741.

3. Tim Watson, *Electrotherapy: Evidence-Based Practice* (Churchill Livingstone, 2008).

4. Ashok Agarwal et al., "Effect of Cell Phone Usage on Semen Analysis in Men Attending Infertility Clinic: An Observational Study," *Fertility and Sterility* 89, no. 1 (January 2008): 124–28, doi:10.1016/j.fertnstert.2007.01.166.

5. Mary Redmayne and Olle Johansson, "Could Myelin Damage from Radiofrequency Electromagnetic Field Exposure Help Explain the Functional Impairment Electrohypersensitivity? A Review of the Evidence," *Journal of Toxicology and Environmental Health. Part B, Critical Reviews* 17, no. 5 (2014): 247–58, doi:10.1080/10937404.2014.923356.

6. Sultan Ayoub Meo et al., "Association of Exposure to Radio-Frequency Electromagnetic Field Radiation (RF-EMFR) Generated by Mobile Phone Base Stations with Glycated Hemoglobin (HbA1c) and Risk of Type 2 Diabetes Mellitus," *International Journal of Environmental Research and Public Health* 12, no. 11 (November 2015): 14519–28, doi:10.3390/ijerph121114519.

7. Howard H. Carter et al., "Cardiovascular Responses to Water Immersion in Humans: Impact on Cerebral Perfusion," *American Journal of Physiology. Regulatory, Integrative and Comparative Physiology* 306, no. 9 (May 2014): R636–640, doi:10.1152/ajpregu.00516.2013.

CHAPTER 12: HEAD STRONG SUPPLEMENTS

1. Florian Koppelstaetter et al., "Influence of Caffeine Excess on Activation Patterns in Verbal Working Memory" (Scientific poster, RSNA Annual Meeting 2005, Chicago, Illinois, December 1, 2005), http://archive.rsna.org/2005/4418422.html.
2. Gabriel S. Chiu et al., "Hypoxia/Reoxygenation Impairs Memory Formation via Adenosine-Dependent Activation of Caspase 1," *Journal of Neuroscience: The Official Journal of the Society for Neuroscience* 32, no. 40 (October 3, 2012): 13945–55, doi:10.1523/JNEUROSCI.0704–12.2012.
3. R. C. Loopstra-Masters et al., "Associations between the Intake of Caffeinated and Decaffeinated Coffee and Measures of Insulin Sensitivity and Beta Cell Function," *Diabetologia* 54, no. 2 (February 2011): 320–28, doi:10.1007/s00125–010–1957–8.
4. Salome A. Rebello et al., "Coffee and Tea Consumption in Relation to Inflammation and Basal Glucose Metabolism in a Multi-Ethnic Asian Population: A Cross-Sectional Study," *Nutrition Journal* 10 (June 2, 2011): 61, doi:10.1186/1475–2891–10–61.
5. Andrew M. James et al., "Interactions of Mitochondria-Targeted and Untargeted Ubiquinones with the Mitochondrial Respiratory Chain and Reactive Oxygen Species," *Journal of Biological Chemistry* 280, no. 22 (June 3, 2005): 21295–312, doi:10.1074/jbc.M501527200.
6. Dana E. King et al., "Dietary Magnesium and C-Reactive Protein Levels," *Journal of the American College of Nutrition* 24, no. 3 (June 2005): 166–71.
7. Kevin A. Feeney et al., "Daily Magnesium Fluxes Regulate Cellular Timekeeping and Energy Balance," *Nature* 532, no. 7599 (April 21, 2016): 375–79, doi:10.1038/nature17407.
8. Sean R. Hosein, "Can Vitamin D Increase Testosterone Concentrations in Men?," *CATIE–Canada's Source for HIV and Hepatitis C Information*, September 2011, http://www.catie.ca/en/treatmentupdate/treatmentupdate-185/nutrition/can-vitamin-increase-testosterone-concentrations-men.
9. Pietro Ameri et al., "Interactions between Vitamin D and IGF-I: From Physiology to Clinical Practice," *Clinical Endocrinology* 79, no. 4 (October 2013): 457–63, doi:10.1111/cen.12268.
10. Akash Sinha, "Shining Some Light on the Powerhouse of the Cell—Is There a Link between Vitamin D and Mitochondrial Function in Humans?" (Conference abstract, Canadian Pediatric Endocrine Group Annual Meeting, Montréal, QC, February 22, 2014).
11. Caroline Rae et al., "Oral Creatine Monohydrate Supplementation Improves Brain Performance: A Double-Blind, Placebo-Controlled, Cross-

over Trial.," *Proceedings of the Royal Society B: Biological Sciences* 270, no. 1529 (October 22, 2003): 2147–50, doi:10.1098/rspb.2003.2492.

12. Alexander M. Wolf et al., "Astaxanthin Protects Mitochondrial Redox State and Functional Integrity against Oxidative Stress," *Journal of Nutritional Biochemistry* 21, no. 5 (May 2010): 381–89, doi:10.1016/j.jnutbio.2009.01.011.
13. U. Justesen, P. Knuthsen, and T. Leth, "Determination of Plant Polyphenols in Danish Foodstuffs by HPLC-UV and LC-MS Detection," *Cancer Letters* 114, no. 1–2 (March 19, 1997): 165–67.
14. http://umm.edu/health/medical/altmed/herb/green-tea.
15. D. O. Kim et al., "Sweet and Sour Cherry Phenolics and Their Protective Effects on Neuronal Cells," *Journal of Agricultural and Food Chemistry* 53 (2005): 9921–7.
16. Tiffany Greco and Gary Fiskum, "Brain Mitochondria from Rats Treated with Sulforaphane Are Resistant to Redox-Regulated Permeability Transition," *Journal of Bioenergetics and Biomembranes* 42, no. 6 (December 2010): 491–97, doi:10.1007/s10863–010–9312–9.
17. J. M. Haslam and H. A. Krebs, "The Permeability of Mitochondria to Oxaloacetate and Malate," *Biochemical Journal* 107, no. 5 (May 1968): 659–67; B. S. Meldrum, "Glutamate as a Neurotransmitter in the Brain: Review of Physiology and Pathology," *Journal of Nutrition* 130, no. 4S Suppl. (April 2000): 1007S–15S.
18. Cameron Rink et al., "Oxygen-Inducible Glutamate Oxaloacetate Transaminase as Protective Switch Transforming Neurotoxic Glutamate to Metabolic Fuel during Acute Ischemic Stroke," *Antioxidants and Redox Signaling* 14, no. 10 (May 15, 2011): 1777–85, doi:10.1089/ars.2011.3930.
19. Francisco Campos et al., "Blood Levels of Glutamate Oxaloacetate Transaminase Are More Strongly Associated with Good Outcome in Acute Ischaemic Stroke Than Glutamate Pyruvate Transaminase Levels," *Clinical Science* 121, no. 1 (July 2011): 11–17, doi:10.1042/CS20100427.
20. M. Yudkoff et al., "Brain Amino Acid Metabolism and Ketosis," *Journal of Neuroscience Research* 66, no. 2 (October 15, 2001): 272–81, doi:10.1002/jnr.1221; John P. M. Wood and Neville N. Osborne, "Zinc and Energy Requirements in Induction of Oxidative Stress to Retinal Pigmented Epithelial Cells," *Neurochemical Research* 28, no. 10 (October 2003): 1525–33.
21. J. D. Johnson, D. J. Creighton, and M. R. Lambert, "Stereochemistry and Function of Oxaloacetate Keto-Enol Tautomerase," *Journal of Biological Chemistry* 261, no. 10 (April 5, 1986): 4535–41.
22. Montserrat Marí et al., "Mitochondrial Glutathione, a Key Survival Antioxidant," *Antioxidants and Redox Signaling* 11, no. 11 (November 2009): 2685–2700, doi:10.1089/ARS.2009.2695.
23. K. A. Bauerly et al., "Pyrroloquinoline Quinone Nutritional Status Alters Lysine Metabolism and Modulates Mitochondrial DNA Content in the

Mouse and Rat," *Biochimica et Biophysica Acta* 1760, no. 11 (November 2006): 1741–48, doi:10.1016/j.bbagen.2006.07.009.

24. Calliandra B. Harris et al., "Dietary Pyrroloquinoline Quinone (PQQ) Alters Indicators of Inflammation and Mitochondrial-Related Metabolism in Human Subjects," *Journal of Nutritional Biochemistry* 24, no. 12 (December 2013): 2076–84, doi:10.1016/j.jnutbio.2013.07.008.
25. Kathryn Bauerly et al., "Altering Pyrroloquinoline Quinone Nutritional Status Modulates Mitochondrial, Lipid, and Energy Metabolism in Rats," *PLOS ONE* 6, no. 7 (July 21, 2011), doi:10.1371/journal.pone.0021779.
26. F. M. Steinberg, M. E. Gershwin, and R. B. Rucker, "Dietary Pyrroloquinoline Quinone: Growth and Immune Response in BALB/C Mice," *Journal of Nutrition* 124, no. 5 (May 1994): 744–53.
27. Kei Ohwada et al., "Pyrroloquinoline Quinone (PQQ) Prevents Cognitive Deficit Caused by Oxidative Stress in Rats," *Journal of Clinical Biochemistry and Nutrition* 42, no. 1 (January 2008): 29–34, doi:10.3164/jcbn.2008005.

CHAPTER 13: BEYOND THE LIMITS

1. M. Costanzo et al., "Low Ozone Concentrations Stimulate Cytoskeletal Organization, Mitochondrial Activity and Nuclear Transcription," *European Journal of Histochemistry* 59, no. 2 (April 21, 2015), doi:10.4081/ejh.2015.2515.
2. Oliver Tucha and Klaus W. Lange, "Effects of Nicotine Chewing Gum on a Real-Life Motor Task: A Kinematic Analysis of Handwriting Movements in Smokers and Non-Smokers," *Psychopharmacology* 173, no. 1–2 (April 2004): 49–56, doi:10.1007/s00213-003-1690-9.
3. R. J. West and M. J. Jarvis, "Effects of Nicotine on Finger Tapping Rate in Non-Smokers," *Pharmacology, Biochemistry, and Behavior* 25, no. 4 (October 1986): 727–31.
4. G. Mancuso et al., "Effects of Nicotine Administered via a Transdermal Delivery System on Vigilance: A Repeated Measure Study," *Psychopharmacology* 142, no. 1 (n.d.): 18–23, doi:10.1007/s002130050857.
5. A. C. Parrott and G. Winder, "Nicotine Chewing Gum (2 Mg, 4 Mg) and Cigarette Smoking: Comparative Effects upon Vigilance and Heart Rate," *Psychopharmacology* 97, no. 2 (1989): 257–61.
6. S. Phillips and P. Fox, "An Investigation into the Effects of Nicotine Gum on Short-Term Memory," *Psychopharmacology* 140, no. 4 (December 1998): 429–33; F. Joseph McClernon, David G. Gilbert, and Robert Radtke, "Effects of Transdermal Nicotine on Lateralized Identification and Memory Interference," *Human Psychopharmacology* 18, no. 5 (July 2003): 339–43, doi:10.1002/hup.488.; D. V. Poltavski and T. Petros, "Effects of Transdermal Nicotine on Prose Memory and Attention in Smokers and Nonsmokers," *Physiology and Behavior* 83, no. 5 (January 17, 2005): 833–43, doi:10.1016/j.physbeh.2004.10.005.
7. Maryka Quik et al., "Chronic Oral Nicotine Normalizes Dopaminergic

Function and Synaptic Plasticity in 1-Methyl-4-Phenyl-1,2,3,6-Tetrahydropyridine-Lesioned Primates," *Journal of Neuroscience* 26, no. 17 (April 26, 2006): 4681–89, doi:10.1523/JNEUROSCI.0215–06.2006.

8. David Nutt et al., "Development of a Rational Scale to Assess the Harm of Drugs of Potential Misuse," *Lancet* 369, no. 9566 (March 2007): 1047–53, doi:10.1016/S0140–6736(07)60464–4.

9. William K. K. Wu and Chi Hin Cho, "The Pharmacological Actions of Nicotine on the Gastrointestinal Tract," *Journal of Pharmacological Sciences* 94, no. 4 (April 2004): 348–58.

10. Rebecca Davis et al., "Nicotine Promotes Tumor Growth and Metastasis in Mouse Models of Lung Cancer," *PLOS ONE* 4, no. 10 (October 20, 2009), doi:10.1371/journal.pone.0007524.

11. Hani Atamna et al., "Methylene Blue Delays Cellular Senescence and Enhances Key Mitochondrial Biochemical Pathways," *FASEB Journal* 22, no. 3 (March 2008): 703–12, doi:10.1096/fj.07–9610com.

12. David J Bonda et al., "Novel Therapeutics for Alzheimer's Disease: An Update," *Current Opinion in Drug Discovery and Development* 13, no. 2 (March 2010): 235–46.

13. Narriman Lee Callaway et al., "Methylene Blue Improves Brain Oxidative Metabolism and Memory Retention in Rats," *Pharmacology, Biochemistry, and Behavior* 77, no. 1 (January 2004): 175–81.

14. Pavel Rodriguez et al., "Multimodal Randomized Functional MR Imaging of the Effects of Methylene Blue in the Human Brain," *Radiology* 281, no. 2 (June 28, 2016): 516–26, doi:10.1148/radiol.2016152893.

15. Hani Atamna and Raj Kumar, "Protective Role of Methylene Blue in Alzheimer's Disease via Mitochondria and Cytochrome c Oxidase," *Journal of Alzheimer's Disease: JAD* 20 Suppl. 2 (2010): S439–452, doi:10.3233/JAD-2010–100414.

16. A. Scott and F. E. Hunter, "Support of Thyroxine-Induced Swelling of Liver Mitochondria by Generation of High Energy Intermediates at Any One of Three Sites in Electron Transport," *Journal of Biological Chemistry* 241, no. 5 (March 10, 1966): 1060–66.

17. Laszlo Vutskits et al., "Adverse Effects of Methylene Blue on the Central Nervous System," *Anesthesiology* 108, no. 4 (April 2008): 684–92, doi:10.1097/ALN.0b013e3181684be4.

18. Murat Oz, Dietrich E. Lorke, and George A. Petroianu, "Methylene Blue and Alzheimer's Disease," *Biochemical Pharmacology* 78, no. 8 (October 15, 2009): 927–32, doi:10.1016/j.bcp.2009.04.034.

19. Uta Keil et al., "Piracetam Improves Mitochondrial Dysfunction Following Oxidative Stress," *British Journal of Pharmacology* 147, no. 2 (January 2006): 199–208, doi:10.1038/sj.bjp.0706459; Kristina Leuner et al., "Improved Mitochondrial Function in Brain Aging and Alzheimer Disease—The New Mechanism of Action of the Old Metabolic Enhancer Piracetam," *Fron-*

tiers in Neuroscience 4 (September 7, 2010), doi:10.3389/fnins.2010.00044; Carola Stockburger et al., "Improvement of Mitochondrial Function and Dynamics by the Metabolic Enhancer Piracetam," *Biochemical Society Transactions* 41, no. 5 (October 2013): 1331–34, doi:10.1042/BST20130054; Rute A. P. Costa et al., "Protective Effects of L-Carnitine and Piracetam against Mitochondrial Permeability Transition and PC3 Cell Necrosis Induced by Simvastatin," *European Journal of Pharmacology* 701, no. 1–3 (February 15, 2013): 82–86, doi:10.1016/j.ejphar.2013.01.001.

20. Ana Latorre-Pellicer et al., "Mitochondrial and Nuclear DNA Matching Shapes Metabolism and Healthy Ageing," *Nature* 535, no. 7613 (July 28, 2016): 561–65, doi:10.1038/nature18618.

INDEX

Page numbers in *italics* refer to charts.

ABOUT THE AUTHOR

Dave Asprey is a Silicon Valley tech entrepreneur, a professional biohacker, a *New York Times* bestselling author, and the creator of Bulletproof Coffee made with butter. He is the host of *Bulletproof Radio*, the Webby Award–winning, number one–ranked podcast, with 50 million downloads, and is actively pursuing a plan to live at least 180 years. He has been featured on the *Today* show, Fox News, *Nightline*, and CNN and in the *Financial Times*, *GQ*, *Men's Fitness*, *Rolling Stone*, *Men's Health*, *Vogue*, *Marie Claire*, the *New York Times*, *Cosmopolitan*, *Forbes*, and dozens more. He lives in Victoria, British Columbia, and Seattle, Washington.